Dedicated to my father,
who devoted his life to our Family,
Aviation and his Country!

Front cover photo - The Air Museum - Planes of Fame Mitsubishi A6M5, an original Japanese Zero-Sen flying near the Southern California coastline at sunset.
Above photo - The Air Museum - Planes of Fame Grumman F3F-2.
Both aircraft were being piloted by John Maloney.
Opposite page - CAL-AERO Academy Unit Insignia.
Back cover photo - an original CAL-AERO Academy pillow case.

Published by - Fox-2 Productions
Design layout by - Joe Cupido
Edited by Sera and Jim Cupido, Denise Porter and many friends

FIRST EDITION
ISBN 0-9701815-0-7
Printed by Dai Nippon, Printing - Japan
Printed in Hong Kong

Any inquires or suggestions should be directed to:

Fox-2 Productions
P.O. Box 20121, Riverside, California 92516 USA
(909) 274-0547 - Fax (909) 274-0548
airfoto @ aol.com
www.fox2productions.com

CHINO

WARBIRD TREASURES
PAST AND PRESENT

FOX-2 PRODUCTIONS
JOE CUPIDO

Big Beautiful Doll
GUN CAMERA

TABLE OF CONTENTS

FOREWARD

The use of ex-military aircraft has gone through many changes since the end of World War II. It wasn't until the late 1940's that the Air Racing arena saw the first application of ex-military fighter aircraft used in another fashion other than fighting in combat. Throughout the late 1950's many civilian owned military aircraft were becoming popular for private recreation while some were being used as business aircraft. During the 1960's, some commercial use of these aircraft were as aerial fire fighting tankers, while others were used by colleges for research projects. Also, many aviation schools were acquiring aircraft for use as training aids. This allowed students hands-on training on real aircraft.

There were many government agencies using ex-military aircraft in a number of different roles such as; VIP support, law enforcement, cargo transportation and research. The early 1960's saw the interest in air racing become popular again. During this time many ex-military aircraft were being sought after, acquired, modified and raced some of which are still flown on the racing circuit.

The late 1950's saw a number of aviation museums around the country starting to acquire ex-military aircraft for display as war treasures. At first, most museums were government owned and operated. Only a few privately owned museums were located around the country, but they were also perserving aircraft. It wasn't until early 1980 that a mass of military bases and civilian aviation collectors started hunting down and saving aircraft.

There are approximately 950 ex-World War II aircraft in museums around the world. This is a very small amount of aircraft, considering that during the war years there were more than 500,000 aircraft produced by all the countries involved in the war. Aviation museums today are packed with artifacts, memorabilia and many types of aircraft used during the war years. Some of these aircraft are the sole surviving example of their type, making them really priceless treasures.

What is a Warbird? There is some disagreement on this subject, but most owners, collectors and enthusiasts have accepted that any aircraft that was designed to use in support of a military role can be classified as a "Warbird." This means, whether these aircraft were fighters, bombers, transports, trainers, or observation aircraft, they could be classified as a Warbird. This would include aircraft used during World War I & II, the Korean Conflict, those used during Vietnam and any aircraft utilized between these era's, to include propeller driven, jet powered airplanes and helicopters.

Currently there are an estimated (excluding training aircraft) 50 – WWI, 600 – WWII, 200 – Korean Conflict, 100 Vietnam era combat aircraft of all types flying in civilian hands in countries around the world. There are also an estimated 200 bomber and fighter type aircraft of all types under different stages of restoration in hangars around the world. The amount of training and support aircraft of all eras can only be guessed at well over 2,500 aircraft.

There is also an unknown amount of aircraft that have not been recovered from their crash sites. Some because of their locations being too hard to get to and others because of the policies of the country where the aircraft lies. There are probably a large number of aircraft that have not even been located on islands, underwater in lakes, or in one of the seven oceans of the world. Many more aircraft have been located in the former Soviet Union and some have been recovered, but because of an unstable economy it has become harder to retrieve these aircraft.

Once these aircraft are located and recovered, their current condition determines if they can be re-built to static display quality or, possibly brought back to flying condition. In either case the process brings back a part of aviation history for future generations to enjoy.

Restoring or rebuilding an aircraft to the condition that it was in when it rolled off the production line requires a great deal of time and money. The process is done by completely overhauling existing parts or manufacturing new ones. A ground up restoration, where every part of the aircraft is stripped, cleaned, rebuilt, or replaced with a new part takes many man-hours and lots of money. The process of bringing an aircraft back to its original flying condition requires even more attention, precise standards, good quality control and craftsmanship. The technology is available today that allows these processes to be completed and warbirds can and are being remanufactured.

The true Warbird movement didn't emerge until the 1970's. Over the years there have been quite a few companies that restored ex-military aircraft for operation in the civilian market. These facilities have been located on airfields all around the world, but none have the history nor the amount facilities are located on Chino Airport, Chino, California.

CAL-AERO Field, or Chino Airport as it is currently known, is located in Southern California and is recognized throughout the world for being the heart of the Warbird community. Chino was one of the first airports to have a heavy concentration of privately owned Warbirds and of Warbird reconstruction facilities. There are currently more Warbirds, museums, reconstruction facilities and Warbird type aircraft on Chino Airport than on any one airfield or airport in the world!

USAF
U.S. AIR FORCE
PILOT

INTRODUCTION

The concept for this book was started over 5 years ago when I realized the importance of Chino Airport to the Warbird community. Over the years I had approached a number of editors / publishers about doing a book on Chino, but none seemed interested. Most felt that there wasn't enough interest in the subject to publish a book on the material. Since I still believed that there was a strong interest, I pursued the project on my own. The following material is an effort to satisfy a dream I have had for over 20 years. That dream is to design and publish a book on my photographic material under my own company's name, "Fox-2 Productions".

During the past 10 years, I had been photographing aircraft at Chino, but it wasn't until about 5 years ago that I started to acquire written material on the subject. The more material I acquired, the more I realized how important the Chino Warbird facilities played in making the worldwide Warbird movement possible.

I would personally like to thank my friend "Uncle" Bob Nightingale for his ideas, information and inspiration, which guided me toward doing this book. I would like to thank all those people that made "Chino" what it was in the past and what it is today. Without their many ideas, dedication and time to the Warbird movement, there would not be a story.

I thank John Maloney and Steve Hinton and many others for their knowledge and experiences. I would also thank all those people that provided their time, cooperation, flying skills, friendship and understanding that helped me acquire the material and made this endeavor possible.

As important, without the photography of Philip Wallick, Jerry Wilkins, Frank Morillo, Carl Porter, Gerald Liang, William Larkins and Emil Strasser, the story would not be complete.

Special thanks to Paul Coggan, John Chapman and Geoff Goodall for their efforts in putting "Warbirds Directory" together -- it became a major source of information.

I would say that given the material I had available, this book is as complete a story about "Chino" as I could make. I am sure that there is much more information that could have and maybe should have been enclosed. The information was just not there when I needed it.

I hope that you find **"CHINO" Warbird Treasures - Past and Present**, an enjoyable, informative and photographically pleasing experience.

U.S. AIR FORCE
ARMAMENT

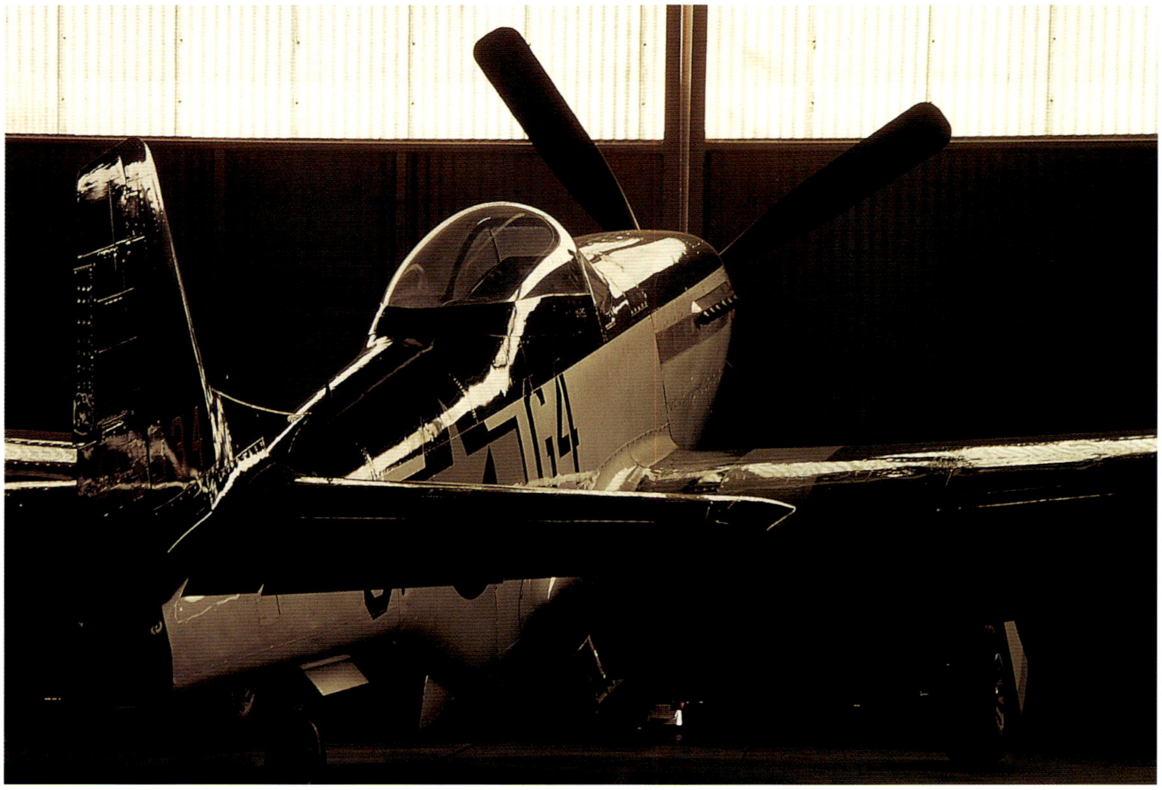

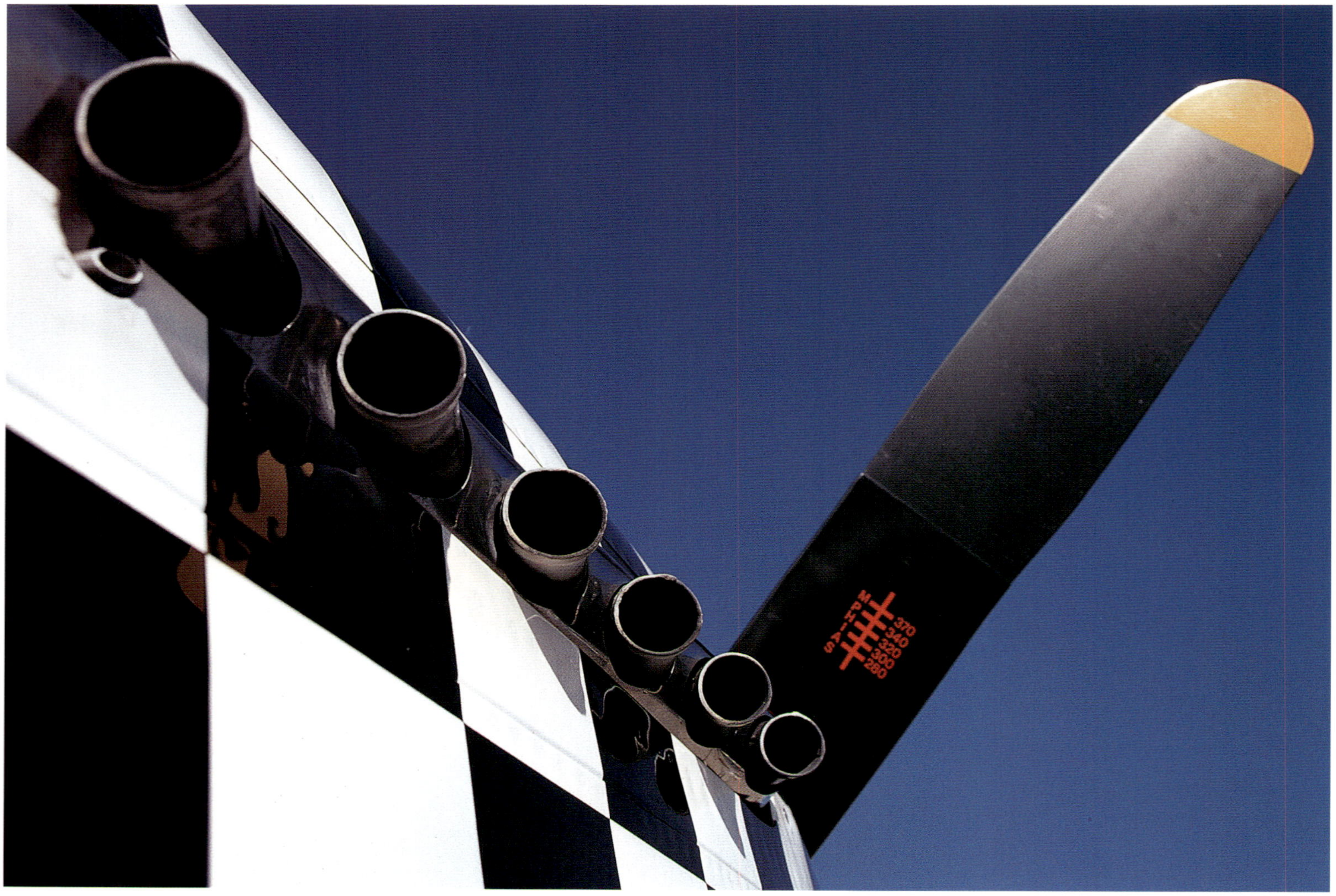
MPH IAS
370
340
320
300
280

The late 1930's saw the United States military in a very vulnerable position with more aircraft than pilots to fly. Prior to the start of World War II, U.S. military pilots and aircrews were being trained at a very slow rate at military airfields around the country. To resolve this problem, General Henry "HAP" Arnold, Chief of the Army Air Corps, decided to employ civilian flight schools to train more pilots and aircrew members.

The U.S. Government awarded contracts to a number of different civilian contractors to open flying schools around the country. By the spring of 1940, 42 flight academies had opened to train more than 30,000 military pilots. These schools were scattered across the country and completely staffed by civilian flight instructors. Most of the flight instructors were highly experienced pilots with over 1,000 flight hours and were also associated with a local U.S. Army Air Corps Reserve Squadron located close to the school where they taught. These schools became a vital part of the war effort training military pilots. They also provided qualified replacement pilots for the normal attrition rate from combat losses.

It just happened that General Arnold's brother-in-law was Major C.C. Moseley who was president of CAL-AERO flight school and CEO of Curtiss-Wright Technical and Aircraft Industries Corporation located in Glendale, California. Major Mosley was also the General Manager of the Grand Central Air Terminal, at the Glendale Airport. Major Mosley was awarded a contract to operate one of the military's new flight schools, which was to be located at Glendale. The problem was that Grand

The opening ceremony at CAL-AERO Academy and Cal-Aero Field in the begining of 1941. There were just 2 hangars and 8 support buildings. By the end of the year 2 more hangars and 8 more buildings were added. 15 - Boeing PT-17 Stearman, 3 - North American AT-6 Texans and a Douglas DC-2 can be seen parked on the ramp in 1941.

The final configuration of CAL-AERO Academy in 1941, 4 hangars and 17 support buildings. You can see the Air Traffic Control Tower just in front of the center of the hangars. Suprisingly, about half of these buildings are still in use today.

Central Field was too small for the facility and a larger airfield was needed. Major Mosley found a perfect location to build a completely new facility just east of Glendale, near the city of Ontario.

Construction on the facility was started within months and operational soon after. At first the facility only had 2 hangars, a control tower and 8 support buildings. The facility was named "CAL-AERO Academy," at Cal-Aero Field, Ontario, California. Major Mosley leased 145 - Boeing PT-17 Stearman from the government to train the pilots. Cal-Aero Field would become the model flight academy for the other 41 schools that would be built throughout the U.S.

CAL-AERO Academy opened in 1940 as a primary and basic flight training facility. The goal was to train military pilots with a high standard of instruction at a lower cost to the government. The cost to turn a cadet into a qualified military aviator was $17.50 per flight hour. An average of 130 flight hours was required for each student, for a total cost of about $2,500.00 to train a single pilot.

Though run by civilians, each facility had a military advisor attached as a liaison officer. CAL-AERO Academy established the standard for flight training programs. The benefit of using civilian flight instructors was that it released the more qualified military instructors for front line combat duties where they were needed.

In the beginning, the program consisted of three 10-week classes. These being, primary, basic and advanced flight training. Before starting to fly, cadets attended ground school where they learned mathematics, weather, navigation, radio operation, aircraft parts and operation, engine operation, aircraft aerodynamics, aircraft performance and some basic understanding of aircraft maintenance procedures.

One of the many Graduation Days at CAL-AERO Academy. Normal class size consisted of 6 flights with an average of 40 students per flight. More than 50 - Vultee BT-13's can be seen parked on the ramp at Cal-Aero Field.

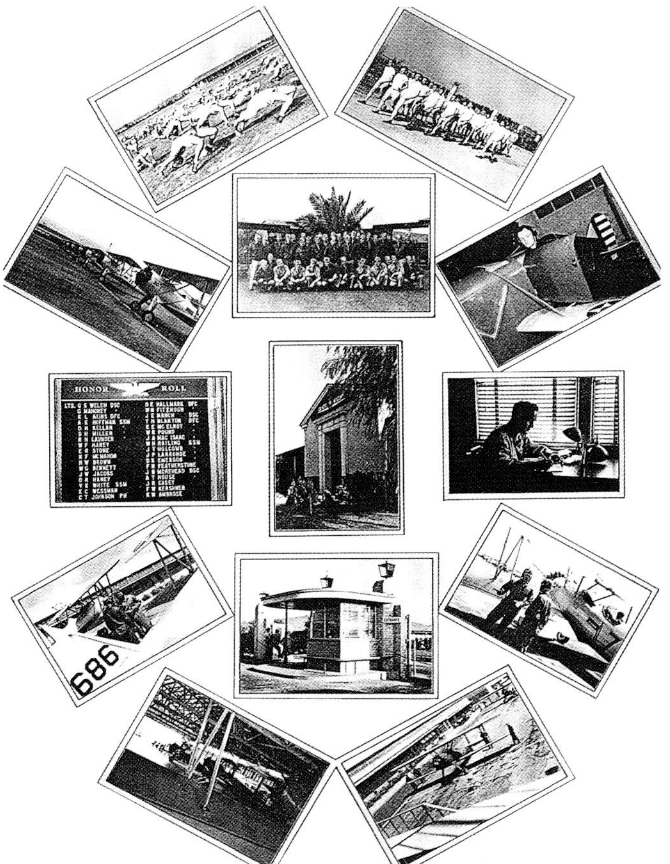

Parked between hangars 2 and 3 at CAL-AERO Academy is a North American NA-36 BC-1, which at the time was the newest trainer acquired by the Army Air Corps in 1941. Two of these aircraft visited CAL-AERO Academy on a good will tour in 1941. This boosted the moral of the students by showing them what type aircraft they would be flying when they advanced through training. You can also see the Main Gate in the background.

Slipstream

Except where noted photos of CAL-AERO Academy were provided by the Chino Library or photo copied from the CAL-AERO Academy monthly magazine "Slipstream," which was published by the Ontario Daily Report, a civilian enterprise. (Photographers and Artist are unknown)

Primary flight training started with about six to ten flight hours before the student pilot would make his solo flight. During the first flights the instructor would teach students to perform take-offs, landings, standard rate turns, slow flight and stalls. Usually, within ten hours a student was capable of solo flight. The next phase of basic flight the student has more solo flying time to become proficient and build their confidence level. More advanced flight maneuvering was also taught; shallow and steep turns, turns around a point, power on and power off stalls and diving the aircraft.

Advanced flight consists of acrobatic maneuvering, snap and slow rolls, chandelles, loops, diving s-turns, inverted flight, Immelmanns, formation flying and instrument flight. The instrument portion of training was accomplished in a link trainer and later in the aircraft under a hood, much like it is performed today.

The academy had taught programs for pilots, bombardiers and navigators. Students entered the program as U.S. Army Air Cadets and upon graduation they were commissioned 2nd Lieutenants. At that point most were sent to either the Pacific or the European Theater to help in the war efforts.

During the first year of operation, CAL-AERO Academy programs were so successful the school was upgraded to teach only basic flight instruction and all primary pilot training was transferred to the "Polaris Flight Academy" in Palmdale, California. This made the training at each school more individualized, meaning that a student would progress from basic flight school at one academy, then transfer to a primary flight training facility, and then finish advanced flight training at yet another academy. The process allowed more students to be trained at one flight academy, rather than having the complete program at one facility. It also allowed facilities to operate more efficiently.

The specialized program allowed one type aircraft to be operated and maintained at each academy, rather than the three that where required to completely train a pilot through all the flight training phases. The Stearman was the primary aircraft used for primary flight school, while BT-9's and BT-13's were used for basic flight school and AT-6's were used for advanced flight training.

Graduating students stand in the front row with lower classmates behind, during the formal event held at CAL-AERO Academy.

It was not uncommon to see a formation flight of military training aircraft flying in Southern California during the war. Here, three Boeing PT-17 Stearman fly together over farmlands near CAL-AERO Field as part of the student's formation training.

The military had the country broken down into sectors, each with a Training Command. CAL-AERO Academy was classified with the civilian operated flight schools in the Western Training Command. The following is a list of the schools that fell under the Western Training Command:

Primary Flight Schools

CAL-AERO Field	located at Ontario, CA
Eagle Field	located at Dos Palos, CA
Gibbs Field	located at Ft Stockton, TX
Hancook	located at Santa Maria, CA
King Field	located at King City, CA
Mira Loma	located at Oxnard, CA
Morton	located at Blythe, CA
Rankin	located at Tulare, CA
Ryan Field	located at Hemet, CA
Sequoia Field	located at Visalia, CA
Thunderbird #1	located at Glendale, AZ
Thunderbird #2	located at Scottsdale, AZ
Tucson	located at Tucson, AZ

Basic Flight Schools

CAL-AERO Field	located at Ontario, CA
Chico	located at Chico, CA
Gardner	located at Taft, CA
Lemoore	located at Lemoore, CA
Merced	located at Merced, CA
Minter Field	located at Bakersfield, CA
Mirana	located at Mirana, AZ
Pecos	located at Pecos, TX
War Eagle	located at Lancaster, CA

Advanced Flight Schools

Douglas	located at Douglas, AZ
La Hunta	located at La Hunta, CO
Luke Field	located at Phoenix, AZ
Marfa	located at Marfa, TX
Roswell	located at Roswell NM
Stockton Field	located at Stockton, CA
Williams	located at Chandler, AZ

A student pilot prepares to taxi for his first solo flight during training at CAL-AERO Academy.

A student and instructor perform a pre-flight inspection on one of the many Stearman's that operated at CAL-AERO Academy. The aircraft is painted in the standard high visibility blue and yellow training colors. Note the Air Traffic Control Tower in the background.

An instructor watches his student flying a Stearman make his first solo flight and landing as part of his primary flight training.

The aerial photo at the right shows CAL-AERO Academy in late 1943. Visible are the four hangars, the living quarters or barracks and other support buildings that made up the facility. Some of these buildings are still being utilized today. The bottom photo was taken only months before the academy was closed in 1944. The original grass runways can also been seen as well as the new paved runways and taxiways that were installed. Another important note are the eleven aircraft revetments located around the airfield. Today, only the revetment located in the bottom center of the photo is left standing on Chino airport. Also note the number of training aircraft parked on the CAL-AERO Academy ramp area.

(Photographer unknown - bottom photo is from the Dennis Trevino collection.)

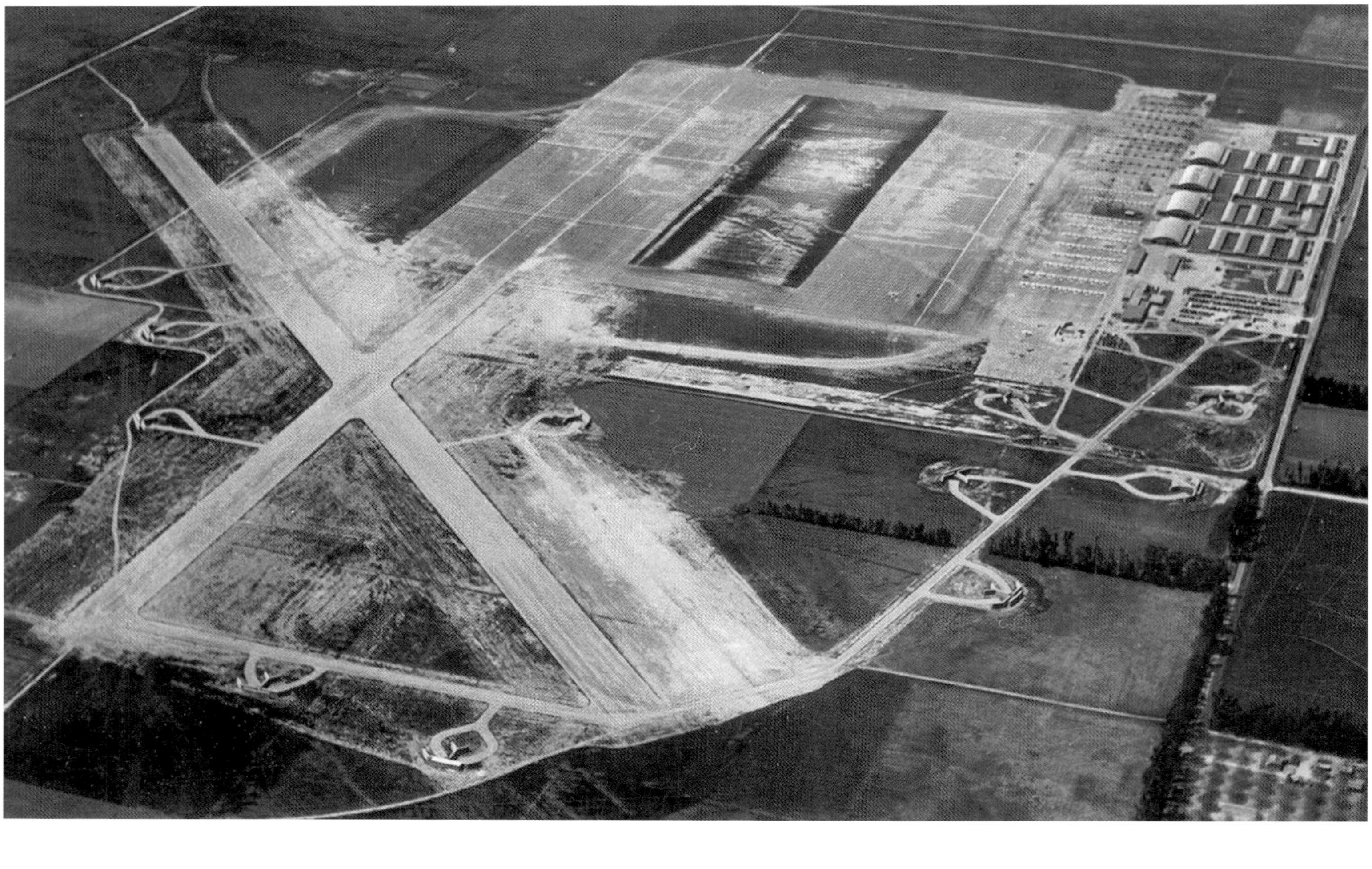

The proof of how effective the training at the civilian flight schools came on the first day of the war. One of CAL-AERO graduates shot down a Japanese aircraft during the attack on Pearl Harbor. Yet, another academy graduate was credited with shooting down four enemy aircraft on his very first combat mission, while flying in the European Theater.

There were eleven CAL-AERO graduates students that flew on the famous "Jimmy Doolittle" raid that bombed the heartland of Japan, only six months after the war started.

Lt John Jerstad, another CAL-AERO graduate was awarded the Congressional Medal of Honor. With his guns jammed or out of ammunition he rammed a Japanese fighter that was attacking his commanding officer and prevented him from being shot down.

Lt Bill Runey, shot down two Lily bombers on the same sortie and later the same year he shot down three Oscar fighters, but was only credited for one kill.

CAL-AERO Academy had the lowest washout rate of all the civilian schools. The pilots that trained at CAL-AERO Academy flew more than 80,000 air miles of training over the four years the school was open. During that period there were only 8 fatalities in all of the civilian schools. This compared to World War I pilot training, when there was one fatality for every 1100 hours of flight training.

One of the many graduates of CAL-AERO Academy was Lt Bill Runey in class 43B. After graduation he was assigned to the 49th Fighter Group / 8th Squadron in New Guinea. Standard flight gear when flying Trans Pacific Island hopping was an oxygen mask, leather helmet, goggles and life preserver. While flying Curtiss P-40's, Lt Runey was credited with shooting down two Japanese Ki-48 Lily bombers on October 15, 1943 and a Ki-43 Oscar fighter later that same year. (Photos courtesy Bill Runey)

THE ONTARIO RECONSTRUCTION FINANCE CORPORATION FACILITY

After the war, the U.S. wasted no time destroying most of the military hardware that they had built to combat the German and Japanese militaries. The U.S. Navy began disposing of their aircraft by dumping them over the side of the aircraft carrier from which they had operated during the war. The Army Air Corps set up large graveyards in the Philippines, Guam and New Guinea where aircraft were bulldozed into deep trenches. The aircraft that made it back to the States, or those that never left the country were being sent to large storage yards located around the U.S..

The government established agencies called Reconstruction Finance Corporation (RFC) to dispose of or sell as surplus military hardware. They determined that 75% of all military aircraft were too hazardous for general aviation use and should be scrapped. Shortly after CAL-AERO Academy was closed in 1944, the airfield was converted into a civilian airport, known as Chino Airport. It would become one of the government's RFC locations.

The mission of the RFC was to dispose of thousands of aircraft that were built and flown during the war and were no longer needed. The government conducted a test program, which showed that scrapping a Consolidated B-24 bomber took 782 man-hours and produced some 32,000 pounds of junk that covered 100 square feet of floor space. A more practical method of disposing of the aircraft was to melt them in very hot furnaces. This process would become the end of the line for many aircraft that had supported the war effort.

The government awarded two different types of contracts to each of the RFC's. One contract was for the selling of aircraft to the private sector, while the other was for the dismantling and melting down of the other aircraft. Part of the contract requirement specified that the RFC had to have all the aircraft cleared off their storage areas in a 9 to 14 month period from when the contract was signed.

The first salvage operation started in late 1945, when brand new Consolidated B-32 Dominator bombers were flown directly from the production line at the Consolidated factory in Fort Worth, Texas to the Walnut Ridge, Arkansas RFC and melted down. In that same year, a total of 23,707 aircraft were melted in furnaces at 1,250 degrees and made into 1,500 pound, aluminum ingots.

North American B-25C-1-NA SN41-13266 was one of 12, B-25 Mitchell medium bombers that was seen at Ontario RFC in 1946 prior to being scrapped.

A rare Curtiss A-25-20-CS, SN42-79765 sitting in the weeds, one of only 120 that where scrapped at the Ontario RFC over the years. The A-25 is the U.S. Army Air Corps version of the Navy's SB2C Helldiver.

Another rare aircraft at the Ontario RFC was this Consolidated OA-10A, SN40-4994, Catalina. A U.S. Army Air Corps version of the Navy's PBY-5A, it is believed to have been the only one of its type to go through Ontario.

There were 5 ex-Navy Grumman J2F-4 Ducks at the Ontario RFC in 1944. The total number of these aircraft that were scrapped is not known.

Only 6 of the 28 RFC facilities were equipped to do the melting process. The others were used only for storage, dismantling, or holding yards for auctions. The 6 that housed the melting pots where: Altus, Oklahoma / Clinton, Oklahoma / Augusta, Georgia / Walnut Ridge, Arkansas / Kingman, Arizona and Ontario, California. Most people think that Kingman was the largest facility, but actually it was the Clinton facility in Oklahoma.

There were more than 35,000 aircraft at these facilities in 1945 waiting to be scrapped. They were distributed in quantities to the facilities as follows; Clinton – 7,600 / Kingman – 5,540 / Walnut Ridge – 4,890 / Augusta – 1,540 / Ontario – 1,390 aircraft. Of this 20,960 aircraft, more than 18,000 were melted down into ingots and recycled. The rest of the aircraft were sold as surplus at auctions.

During the beginning of 1946 there were a total of 33,983 aircraft at the RFC facilities around the country. Of that number 23,707 were melted down into aluminum ingots. By the end of the same year, the Surplus War Aircraft Division had sold 10,276 aircraft at auctions at one of the 30 sales depots around the country. All that was needed to ferry aircraft off government property was a private pilot's license. Otherwise, the aircraft had to be dismantled and removed within a period of time after the sale.

Sharp & Fellows Contracting was a Los Angeles based company that received a contract for $404,593.00 to operate the Ontario RFC. The auctions were open to the public and all bidding for aircraft were sealed bids. The first auction sold 489 aircraft of a variety of types. These included: 1 Curtiss C-46 Commando / 288 Cessna UC-78 Bobcats / 16 Douglas C-47 Gooneybirds / 1 Douglas C-54 Skymaster / 28 Boeing PT-17 Stearman and 155 North American AT-6 Texans. Soon after the first auction, the remaining aircraft were scrapped at an alarming rate. There where only a few other aircraft sold to the public before the Ontario RFC closed.

ONTARIO RFC DURING MAY OF 1946

Looking North, you can see the four main hangars and the control tower on the main ramp area of Cal-Aero Field. Also visible are rows of Bell P-39's, Lockheed P-38's, Curtiss P-40's, Douglas A-20's and P-70's, Consolidated B-24's, a few Douglas C-54's and C-47's are parked on the main ramp area. There are also a number of North American T-6's waiting for the cutting torch and melting pot.

Looking West, are numerous practically stripped Bell P-39's, Douglas A-20's, Consolidated B-24's, a few Douglas SBD's or Army Air Corps A-24's, Curtiss A-25's and 2 Lockheeed AT-18's. Most of these aircraft had their engines and landing gear removed and anything else that was useful. The remains awaited the melting pot. There are quite a few rows of Curtiss C-46 "Commandos" that still look totally in tact. This view also shows the relationship of how big the storage facilty was and the quanity of aircraft that were stored and later scrapped at Ontario RFC.

Looking East, a few North American B-25's and rows of Martin B-26's are being disassembled. There are also some rows of Boeing B-17's and a variety of fighter type aircraft in the back. All reusable parts were removed from the aircraft before they were melted down into ingots. Some of the reusable parts included engines, propellers, landing gear, instruments, weapons, radios and avionics. All aircraft plexi-glass was removed and destroyed separately.

Looking South, rows of engines and propellers are in the foreground, while rows of Curtiss C-46's and B-17's, B-24's. There are also four Grumman J2F-4 Ducks and a Consolidated AO-10 Catalina set off to the left. Located in the left lower corner are a North American B-25, a Bell P-39 and Curtiss P-40 stripped and upside down. The hardened revetments were constructed for Lockheed P-38's to sit coastal air defense alert during the war, but they were never used.

Many different types of aircraft were scrapped in the short 14 month period that the Ontario RFC was operational. The following is a list of most of them:

- Bell P-39Q & TP Airacobra
- Boeing B-17E, G Flying Fortress
- Boeing PT-17 Stearman
- Cessna UC-78 / AT-17 Bobcat
- Consolidated B-24D, E, H, J Liberator
- Consolidated PBY-5A / OA-10 Catalina
- Consolidated L-5 Sentinel
- Curtiss C-46A, E Commando
- Curtiss P-40E, F, K, M Kittyhawks
- Curtiss A-25A Helldiver
- Douglas A-20G / P-70 Havoc
- Douglas B-18A Bolo Bomber
- Douglas C-47 Gooneybird
- Douglas C-54 Skymaster
- Fairchild AT-21
- Grumman J2F Duck
- Lockheed P-38F, H, L, Q Lightning
- Lockheed R50-6
- Lockheed AT-18 Hudson
- Lockheed B-37 Ventura
- Martin AT-23A
- Martin B-26B, C Marauder
- North American B-25C, D Mitchell
- North American O-47
- North American P-51A, B, C, D Mustang
- North American F-6 Mustang
- Northrop P-61 Black Widow
- Republic P-43A Lancer
- Republic P-47C, D, G Thunderbolt
- Stinson L-1 Vigilant
- Vought OS2U-3 Kingfisher
- Waco XCG-15A

Imagine what a collection one of each of these aircraft would have made! When the Ontario RFC finally closed, a total of 1,879 aircraft had passed through the facility. Of those, 1,376 were scrapped and 503 were sold at auction. The Ontario facility was turned back over to the city of Chino and it became Chino Airport.

A small group of Douglas P-70B's, the Night Fighter version of the A-20 Havoc, were dismantled and cut up. One of them was SN42-54053.

A total of 120 Martin B-26 Marauders were scrapped at Ontario RFC. Above, B-26B SN41-32031 waits its turn for the cutting torch.

A Lockheed AT-18 Hudson, SN41-2331 sits in the grass with a few Douglas P-70's in the background.

A Consolidated RB-24E, sits alone on the West ramp, the Serial number was unknown. The 341 was a RFC tracking number.

Republic P-47, P-47C-5-RE, SN41-6591 Thunderbolt sits in the tall grass with a few other "P-47's and the only Northrop P-61 Black Widow that was seen at Ontario in 1946.

A total of 140 Consolidated B-24 Libarators were scrapped at Ontario, some of them had very unusual modifications. Shown below is a B-24H-1-FO, SN42734, note the nose faring looks like it might have housed some sort of radar system.

Lockheed P-38G-15-LO, SN43-23359 was the only one of 15 Lightnings that carried the external fuel tanks. Also note the Army Air Corps Douglas A-24 in the background.

Curtiss P-40K-1-CU, SN42-46216 Warhawk sits in the weeds, hoping it will be sold instead of being scrapped. The large faded numbers X845A on the fuselage were used in some way to track the aircraft while going through the RFC. Note the P-39's in the background.

Aircraft	Price
◆ Bell P-39 Airacobra	$ 1,000.00
◆ Bell P-63 Kingcobra	1,000.00
◆ Boeing PT-17 Stearman	1,750.00
◆ Boeing B-17 Flying Fortress	13,750.00
◆ Boeing B-29 Superfortress	32,500.00
◆ Consolidates B-24 Liberator	13,750.00
◆ Consolidated PBY Catalina	13,750.00
◆ Consolidates L-5 Sentinel	1,250.00
◆ Curtiss C-46 Commando	12,250.00
◆ Curtiss P-40 Warhawk	1,250.00
◆ Curtiss A-25 Helldiver	2,000.00
◆ Douglas C-47 Gooneybird	32,500.00
◆ Douglas C-54 Skymaster	85,000.00
◆ Douglas A-20 Havoc	3,000.00
◆ Douglas A-24 Dauntless	2,000.00
◆ Grumman F6F Hellcat	3,500.00
◆ Grumman FM2 Wildcat	1,250.00
◆ Lockheed P-38 Lightning	1,250.00
◆ Lockheed AT-18 Hudson	2,000.00
◆ Martin B-26 Marauder	3,000.00
◆ North American T-6 Texan	500.00
◆ North American P-51 Mustang	3,500.00
◆ North American B-25 Mitchell	7,250.00
◆ North American O-47	1,250.00
◆ Northrop P-61 Black Widow	6,000.00
◆ Republic P-47 Thunderbolt	3,500.00
◆ Vought F4U Corsair	1,250.00
◆ Vultee BT-13	450.00

The government priced aircraft on ease of conversion for use in the civilian market. This was the reason that transports and heavy bombers were priced so much higher than the fighter aircraft. Cargo aircraft were the most sought after because they would have the most use in a non-military environment, so they yielded the highest dollars. At the time, no one thought that there was a use for fighter aircraft in the civilian market, so those aircraft were the first to be scrapped.

So, by now you're thinking, "How many and which type aircraft would I have bought." The cost of these aircraft is relative to the money that was available to most people in those days. Most families were having a hard enough time just putting food on the table. If the aircraft did not sell at auction, they were scrapped with those already selected for destruction. Very few, if any, of the aircraft were saved.

All photographs in Part I of the aircraft at Ontario RFC, where taken by William T. Larkins in May 1946. They were taken with a Graflex "Press Camera" on Kodak Vigilant film. Aerial photos were taken from a Piper J-3 Cub. This historical documentation is truly priceless.

Boeing B-17F "The Devil Daughter" SN42-23153 showing 26 bombing missions. A total of 12,731 B-17's where built by Boeing, Douglas and Lockheed aircraft companies during its production run. Peak production was reached in 1944 with 16 aircraft being completed every 24 hours.

Ontario RFC scrapped a number of rare Republic P-43A Lancers. Above, SN40-2894 was one of only 272 P-43's produced for the Army Air Corps. The "Lancers" that the Army Air Corps were flown as advanced trainers, while a number of the fighter versions were sent to China and flown by volunteers. The Royal Australian Air Force utilized the reconnaissance version of the Lancer.

Another very rare aircraft that passed through Ontario RFC was this Stinson L-1 Vigilant, SN41-19039. There were 324 L-1's built for use as spotter, observation, air ambulance and liaison duties.

A Lockheed B-37 Ventura, SN41-36341 sits in the weeds. This was one of only a few flown by the Army Air Corps. The U.S. Navy purchased the more advanced "PV" version of the Ventura and used it as a medium patrol bomber.

A large number of Martin B-26 Marauders were scrapped at Ontario. SN41-32061 sits out in the weeds by itself. In the background you can see rows of Douglas C-47's lined up on the main airport ramp area.

A Lockheed AT-18, SN41-23475 sits with other Hudsons waiting to be stripped before going to the melting pot. Note where the rear gun turret was located and has been covered by sheet metal.

An early B-24D, SN42-63893 sits in the same area as the B-26 shown above along with some of the other fighters shown. It is not known if this area was being used to save at least one of each type aircraft. There were a number of complete airframes stored in this same area.

Piper's J-3 Cub served the military well during the war, flying liaison, observation and air ambulance duties. This J-3 Cub SN45-5192 is believed to be the same aircraft that William Larkins used to take the aerial photographs of RFC in 1946.

This Waco XCG-15A, SN44-90986, twin R-755-9 engine powered glider was the only one ever built. This aircraft should have been saved for a museum, but like so many, it was scrapped and is now just a piece of history.

Still wearing the famous "Flying Tiger" style shark mouth nose art Curtiss P-40F sits waiting for the melting pot. Also unique to this photo is a very rare Bell TP-39 dual cockpit Airacobra sitting in the background.

North American built 310, P-51 "A" model Mustangs shown is SN42-83736 sitting with a group of Vultee BT-13's, all waiting to be dismantled and melted down.

Republic P-43B, SN40-2894 Lancer parked next to some stripped Bell P-39's hulks. Because of poor performance, the production run was cut short on the P-43. The aircraft wasn't used very long by the Army Air Corps and only 51 of the intended 125 P-43's were delivered to China.

A number of different versions of the consolidated B-24 Liberator went through Ontario RFC. Above, a "J" model SN42-64371 sits parked on the hard stand, probably soon after it arrived at Ontario. In the background is the only Northrop P-61 Black Widow that was seen at Ontario.

Still wearing its shark-mouth nose art, a Douglas A-20G, SN42-53797 Havoc sits in front of the rare P-70 night fighter version of the A-20. Note that with the six 50 caliber guns removed, the gun ports have been covered with sheet metal. A total of 154 A-20's and P-70's went through Ontario RFC.

Yet another rare aircraft that saw their last days at the Ontario RFC, were a small batch of Fairchild AT-21's. Above, SN42-11698 was one of only 175 produced during early 1940 for use as aerial gunnery and navagator training. What made it unique was that the aircraft was constructed completely of wood and non-vital war materials.

Only a few Douglas B-18A Bolo, went through Ontario. Below, SN37-501 sits waiting its turn. You can see the design features are much like the DC-3 or C-47. The Army Air Coprs operated 370, B-18's as medium bombers before and during the early years of the war. The Royal Canadian Air Force operated 20 B-18's in anti-submarine warfare and reconnaissance duties.

One can only wonder why someone did not have the foresight to save more aircraft than the few that where put away. It would take more than thirty years before military aircraft would be sought for display as historical relics. The military had already established large storage yards on vast areas of land where some of these aircraft could have been saved and preserved! Instead they were destroyed and are now extinct!

One of the last missions all these war machines made was in the civilian aviation community. All the technological advancements in aircraft design and performance made during this period would carry over into the design of both large and small civilian aircraft in the following years. This was one of the most important eras in aviation!

All the photographs in Part II of the Ontario RFC were taken by Emil Strasser on two visits to the facility during 1946. Enough can't be said of the historical value of these photographs. The images were acquired from Gerald Liang's private collection. We are lucky enough to have had William T. Larkins and Emil Strasser with the interest and foresight to photograph and record this time in history!

Above, a Lockheed P-38H, SN42-66948, Curtiss P-40E, had serial number and identification markings painted over. Below, a Bell P-39Q, SN42-20653 were parked in the weeds in an area located at the Southwest corner of the airfield, well away from the masses of aircraft that were being scrapped. All the aircraft parked in this area appeared to be completely intact. Is it possible that there were part of a group of aircraft that were saved? Where are they now?

3 GONE, BUT NOT FORGOTTEN

Following the closure of the Ontario RFC, the airfield was turned over to the city of Chino and a number of privately owned companies moved in and out of the facilities that had been constructed during the war. This chapter covers most of the companies that helped establish the Warbird community at Chino, and worldwide. Though some of the information is a little vague, it is all I could find. It does however, give you an idea of the quantities and variety of companies that have called Chino Airport their home. Some of these companies occupied existing hangars while others built new ones. Chino Airport has continued to expand and improve for the future.

PACIFIC AEROMOTIVE

Pacific Aeromotive Corporation (PAC) was made up of seven divisions located around the country, with the home office based in Burbank, California. The facility at Chino Airport was used to complete government contracts to overhaul military aircraft.

The military saved money by contracting out this type of maintenance to civilian contractors who were able to do the work more efficiently. PAC occupied the 80,000 square feet of the four hangars and most of the support buildings that CAL-AERO Academy had utilized. They also expanded and paved the ramp area to cover 42 acres and added a second runway. PAC operated at Chino Airport from the early 1950's to late 1960's.

The PAC facilities where capable of completely rebuilding aircraft and they operated a number of different maintenance shops. These included a hydraulic shop, instrument, engine shop, electrical shop, propeller shop, fabric and sheet metal fabrication shops and a fully enclosed paint booth.

PAC was contracted to overhaul large and medium transport aircraft, though, toward the end they also rebuilt a number for North American F-86's that were exported. Douglas C-47's and C-54's were the primary aircraft being worked over at

The two photos of the Pacific Aeromotive Corporation facility at Chino show the many airport facilitiy changes and the large number of aircraft that were being worked on at the PAC facility. The photo on the opposite page shows the early years, when almost all of the CAL-AERO Acedemy buildings where utilized by PAC. It also shows the new maintenance shops that where constructed at the east end of the hangars. "Fighter Rebuilders" still operates out of what used to be PAC's engine overhaul shop. The photo below shows how more than half of CAL-AREO Academy support buildings have been removed. It also shows how large of an operation PAC had at Chino over the years. This same photo shows 19 military Douglas C-47 Gooneybirds and a mix of 26 military and civilian Douglas C-54 Skymaster / DC-4 Airliners. You can also see the additonal runway that was added as well as the hardened revetments that where constructed for Lockheed P-38's to sit ready for air defense alert during the war. (Photogapher unknown)

the Chino PAC facility. PAC had facilities all around the country located close to military bases to make it even more efficient for the government to award the contracts to civilian operators.

One of the reasons that PAC located their facility at the Chino Airport was because of its location close to two Air Force bases. Both March Air Force Base (AFB) in nearby Riverside and Norton AFB in San Bernardino, operated aircraft that were being overhauled at PAC. PAC also overhauled military aircraft based at; Travis, Mather, McCellan, Beale, Castle, Hamilton and Edwards, all Air Force Bases located in California.

The PAC Chino facility was a Civil Aviation Administration (CAA), now called the Federal Aviation Administration (FAA) certified repair station. This allowed civilian and airline companies to utilize the services of PAC. TWA, Pan Am, American, and Pacific Northwest all had their aircraft overhauled by PAC, it was just more economical than operating their own overhaul facility.

During the peak years of operation at Chino, PAC employed more than 1,400 workers and overhauled or had major repair work performed on more than 1,300 aircraft. One of the more unique aircraft to pass through PAC for maintenance, was a Portuguese Naval SB-17G Flying Fortress.

PAC closed its Chino based operation after government contracts ended in the early 1960's. This was also about the time that propeller type aircraft operation was replaced by jet transports in both the military and in the airline industry.

AERO-SPORT

Aero Sport operated out of what is known as hangar #2, but this was actually hangar #1 when the CAL-AERO Academy operated from it at Cal-Aero Field, or what is now known as Chino Airport. Aero Sport is well known for promoting the use and the rebuilding of ex-military Warbirds as civilian operated personal aircraft. The founders of Aero Sport were Dick Harriston, J.D. (Soup) Hosington, Jerry Morrison, and later Jack Lee and Dudley Hutchins would buy into the company. These where the grandfathers of the Warbird movement at Chino!

It was in the early 1960's that Aero Sport opened its doors. Soup, Jerry and Jack all believed that ex-military aircraft could play an important role in the civilian market place. The beginning years at "Aero Sport", saw mostly North American P-51 Mustang's being converted for civilian use. This would become the trademark of Aero Sport. Later on, North American T-6 and Harvards were also restored at Aero Sport.

Every now and then odd aircraft received maintenance or an annual inspection at Aero Sport. Some of them were: a Douglas B-26 Invader, a couple of Grumman F8F Bearcats, quite a few North American T-28s an a Grumman SA-16 Albatross.

Jerry Morrison and Dick Harriston, two of the founders of Aero Sport shown working on two Merlin engines. (Photographer unknown)

Aero Sport's operated at Chino between 1962 and 1985. In this period more than twenty P-51 Mustangs were restored or completely rebuilt from wrecked hulks, while many other warbirds received varing degrees of repair and maintenance. Most of these aircraft are still flying today, thanks to Aero Sport!

During normal daily operations at Aero Sport Mustangs would taxi up to the hangar for all types of maintenance, some having minor maintenance, while others were being completely rebuilt. SN 44-73079 / N576GF was parked outside the Aero Sport hangar for minor maintenance. The aircraft was owned and operated by Al Ruggeri, and later owned by Bob Love, well known Korean Fighter Ace. The aircraft now flies as N151BL. In 1973 the aircraft was painted red with white and black stripes. (Photographer unknown)

"Uncle Bob" Nightingale shown working on one
of the many P-51's that passed through Hanger
#2, where Aero Sport operated for many years.
Normal operations saw two or three Mustangs
in or around the Aero Sport hangar.
(Photographer unkown)

Mustangs often arrived dismantled on trailers.
This was because most were non-fliers and it
was the only way to get them to Chino. SN44-
74829 / N169MD came in for a complete restoration during 1973. (Photo by Gerald Liang)

Shown on the ramp near Aero Sport in 1973 is P-51D SN45-11546 / N51JW which was in for minor
maintenance and owned by John Wright. P-51 SN45-11483 / N4674V also in for maintenance, this
aircraft was owned by Dwight Gibson of Los Angeles. This same aircraft also flew as N7096V when
owned by Leo Pike of Bakersfield, California. The North American T-28A SN49-1619 / N7491C was
owned by Richard Holland of Fountain Valley, California and like the Mustangs was in for scheduled
maintenance at Aero Sport. (Photo by Gerald Liang)

Most of the Mustangs still flying today have very unique histories, P-51D SN44-74425 / NL11T is no exception. The aircraft served with the U.S. Army Air Corps then was transferred to the Royal Canadian Air Force and then had 29 private owners. The aircraft currently operates in Holland, where it is operated in the markings of SN44-74425, OC G "Damn Yankee."

After being retired from military service it was registered as N6522D and N51HB, then back to N6522D, before it became NL11T. Early on, NL11T sported a post military silver paint scheme. Shown taxiing into the Aero Sport ramp area between hangar #2 and #3 where Ruth Johnson's Flying Service operated. (Photo by Gerald Liang)

The first P-51D Mustang rebuilt at Aero Sport was in 1963, for Cliff Cummins, P-51D, SN44-15651 / N79111 "Galloping Ghost." The aircraft was later converted to # 69 "Miss Candace," which was raced at Reno for many years. The last Mustang built at Aero Sport was for Rodney Barnes P-51D, SN44-73129 / N51SL in 1985. Elmer Ward had his first P-51D rebuilt at Aero Sport, SN44-72739 / N44727 "Man O'War."

Aero Sport also worked on a variety of other Warbirds. One of the more exotic Mustangs that came through the Aero Sport hangar, was an Ex-Israeli Air Force Mustang that was purchased by a private owner. In the late 1960's Aero Sport also rebuilt three Mustangs for the Bolivian Air Force, two D's and one dual controlled TF-51. These aircraft were completely military stock to include the guns and weapons racks on the wings, though they were not added until the aircraft reached their own country.

Aero Sport also maintained the U.S. ARMY P-51D's, SN44-72990 / N62322T, currently dis-

played at the U.S. Army Aviation Museum, Fort Rucker, Alabama and SN44-13571 / 68-15796 which was a "D" model later converted to a Cavalier Mustang and is currently displayed at the Edwards Air Force Base Museum, Edwards, California. These aircraft were normally based at Edwards Air Force Base and used as chase aircraft for many test programs during the late 1960's and into the early 1970's.

Aero Sport also helped maintain the Fire Fighting aircraft that operated at Chino. Though TBM, Inc. pilot and flight crews were A&P qualified maintenance personal, Aero Sport was called on many times to provide maintenance support to keep the TBM's, F7F's and B-17's operating at the Chino, tanker base.

During the mid-1970's Aero Sport converted an Ex-Coast Guard Grumman SA-16 Albatross for NASA. The aircraft was equipped to carry an experimental side scanning radar which NASA was developing for commercial use.

SN44-63865 / N-51JK is one of the more historically important P-51's still flying. The aircraft was credited with three kills while flying with the 353rd Fighter Squadron in Europe during the war. The aircraft also served with the Swedish and Nicaraguan Air Forces and was raced as #47 at the Reno National Air Races.   (Photo by Gerald Liang)

One of the first military style paint schemes put on a civilian Mustang, "Stump Jumper" SN44-63810. Ed Maloney was the first owner after its retirement from the military. (Photo by Gerald Liang)

One of the most colorful in the 1970's was this bright red Mustang / N5436V maintained at Aero Sport for more than five years. The aircraft is truly a Chino Mustang. Though the owners changed over the years, the aircraft ended back at Chino when the current owner, Bob Pond, had it rebuilt by Figther Rebuilders in 1996. (Photo by Gerald Liang)

"Miss Candace" was maintained and modified at Aero Sport, when she was not being raced. SN44-15651 / N79111 was flown at the National Cleveland and Reno Air Races under a variety of race numbers, making it the most historically raced P-51 currently still flying. (Photo by Gerald Liang)

Following is a list of North American P-51 Mustangs that were rebuilt, maintained or had some sort of maintenance at Aero Sport:

44-13571 / 68-15796 - operated as chase aircraft by the U.S. Army at Edwards Air Force Base, currently displayed at the Edwards Air Force Base Museum, Edwards, California.

44-15651 / N79111 - was the first P-51 built at Aero Sport and was operated by Cliff Cummins as "Galloping Ghost," and later as "Miss Candace," "Jeannie" and "Spectre" and is currently raced as #9 "Cloud Dancer." The aircraft is now owned by Bahia Oaks, Inc, Ocala, Florida.

44-63476 / N51DH - built for Consolidated Airways, Fort Wayne, Indiana, the aircraft is currently operated by The Evergreen Heritage Collection in Oregon.

44-73343 / N5482V - the aircraft crashed and was a total loss soon after being rebuilt for Bruce Morehouse, San Antonio, Texas.

44-63810 / N63810 - originally rebuilt for Planes of Fame Museum, Chino, California, the aircraft is now operated as N451BC "Angels Playmate" owned by Joe Newsome, Cheraw, South Carolina.

44-72739 / N44727 - originally built for and still owned and operated by Elmer and his son Todd Ward, Chino, California as "Man O War."

44-72826 / N6344T - during Feburary of 1976 this aircraft had an annual inspection at Aero Sport, it is currently flying as N51YS, owned by Steve Collins, Dunwoody, Georgia.

44-72990 / N62322T - operated as a chase aircraft for test flights by the U.S.Army at Edwards Air Force Base, currently displayed at U.S. Army Aviation Museum at Fort Rucker, Alabama.

44-73027 / N5747 - built for Mustang Pilots Club, Hollywood, California. After a gear up landing at Mojave, the aircraft was repaired and crashed again gear up, was rebuilt and crashed again, then rebuilt and raced as #56. It is currently flying as F-AZJM "Temptation" and is operated by the Flying Legends Association, Dijon-Longvic, France.

Aero Sport maintained the U.S. Army P-51's stationed at Edwards Air Force Base, California. One of the aircraft was a Cavalier converted P-51D SN44-13571 / SN 68-15796 which was modified with wing tip tanks for extended flight duties during test flight chase missions. The cockpit was configured to carry a passenger or a test flight observer and the canopy was an enlarged version for better visibility. (Photo by Gerald Liang)

Parked between Aero Sport and Ruth Johnson's Flying Service P-51D SN45-11553 / N51T, one of the Mustangs that was in for minor maintenance at Aero Sport. Flo's Restaurant can be seen in the background. (Photo by Gerald Liang)

44-73029 / N79999A - the aircraft is currently flying as N51JB "Bald Eagle" and was raced as #51. The aircraft is owned and operated by James Beasley, Philadelphia, Pennsylvania.

44-73129 / N51SL - was the last P-51 Mustang built at Aero Sport and was owned by Rodney Bar, of Louisville, Kentucky. It crashed on take-off and was stricken from the F.A.A. record book in 1987.

44-73149 / N6340T - the aircraft was raced as #7 "Candy Man" and currently flies as "Moose" with the Fighter Collection, Duxford, England.

44-73206 /N3751D - flew as "Hurry Home Honey," owned by Charles Osbornes. The current owner Stuart Ederhardt of Danville, California, flies N151SE and has raced it as #22 "Merlins Magic."

44-73415 / N6526D - after being rebuilt at Chino and raced as #45 & #55 "Pegasus," the aircraft was sold and is currently raced as #55, "Voodoo Chile," and owned by Delbert Williams.

44-73832 / N8373D - after being rebuilt and flown as N117E by Harvey Wellington, Wellington, Nevada.   The aircraft was stricken from the F.A.A. record book after it crashed in 1992.

One of the two North American P-51D's that the U.S. Army operated as chase aircraft for advanced aircraft flight testing at Edwards Air Force Base, California in the late 1960's and early 1970's. Aero Sport provided both military and civilian contracted maintenance on both Mustangs'. P-51D, SN44-72990 / N6322T / 0-72990 was operated by a number of different owners before the U.S. Army acquired the aircraft; the U.S. Army Air Corps, Royal Canadian Air Force, U.S. Air National Guard and it was being operated by a civilian owner when the U.S Army purchased it. (Photo by Carl Porter)

P-51D, SN44-63810 / N63810 was originally acquired by Ed Maloney at a government surplus sale at Norton Air Force Base, California in 1963. Robin Collard of Merced, California bought it and had it rebuilt at Aero Sport as "Stump Jumper" in 1975. It is currently owned by Joseph Newsome, Cheraw, South Carolina and flies as N451BC "Angels Playmate." (Photo by Gerald Liang)

44-73973 / N51JC - was built for Jerry Janes, of Vancover, British Columbia in 1979. It was later sold to David Price of Santa Monica, California as N51DP and raced as #49 "Cotton Mouth" and is still owned by David Price, Museum of Flying, Santa Monica, California.

44-74008 / N8676E - has a history of being raced as #51 & #71. The aircraft crashed in 1975 and was rebuilt as N151MC. It is currently is owned by Ronald Runyan, Fairfield, Ohio.

44-74425 / NL11T - rebuilt in 1973, over the years the aircraft has been owned by numerous people, since 1994, the aircraft has been flying as SN47-4425 "Damn Yankee" in Holland.

44-74469 / N7723C - restored in 1987, owned by Jerry Miles, Classic Air Parts Inc, Miami, Florida, the aircraft crashed in 1992 and has been stricken from the F.A.A. record books.

44-74756 / N69QF - was owned and flown by Ken Burnstine of Fort Lauderdale, Florida. The aircraft and pilot where lost in an accident in 1976.

P-51D SN44-72826 / N6344T "Danny Boy," shown at Aero Sport during a annual inspection in 1978. The aircraft was sold soon after it left Aero Sport. In its early days, this Mustang flew with the RCAF as 9563 and many years later returned to Canada and operated with the Canadian Warplane Heritage. It currently flies as SN44-72826 / N51YS "Old Boy." (Photo by Gerald Liang)

P-51D SN44-84952 / N210D shown parked by Aero Sport for minor maintenance. The aircraft was a Chino based Mustang owned by Joseph Hartney from 1973 through 1976. (Photo by Gerald Liang)

44-74829 / N169MD - after a crash landing the aircraft was stored for ten years and then rebuilt in 1979. The aircraft is currently flying as ZK-TAF / NZ2415 with the New Zealand Historical Aircraft Trust, Admore, New Zealand.

44-84952 / N210D - aircraft was based at Chino and maintained by Aero Sport between 1973-1976. The aircraft is still registered as N210D and owned by Northwestern Aircraft Associates, Wilmington, Delaware.

44-84753 / N5436V - maintained at Aero Sport between 1973-1978, aircraft is currently owned by Bob Pond and flies as N251BP. It is displayed and operated by the Palm Springs Air Museum, Palm Springs, California.

44-84961 / N7715C - bought as surplus from McCellan Air Force Base, California. It was raced as"Miss RJ," then modified and raced as "Roto Finish #5 and later highly modified in to the "The Red Baron #5 RB-51. The aircraft was lost while racing at the Reno National Air Race in 1979.

44-84962 / N9857P - this aircraft had a history of flying with the Korean and Indonesian Air Forces. It was rebuilt in 1983 and is owned by Lee Schaller, Montville, New Jersey.

45-11483 / N4674V - this aircraft was operated by American Aeronautics, Corporation, Burbank,

California in the 1950's and was sold to Sterling Aircraft and later Leo Pike, who had it rebuilt at Aero Sport. The aircraft is now registered as N7096V.

45-11540 / N5162V - rebuilt in 1974 as N5162V, it currently flies as SN45-11540 / N151W, "Queen B" and is owned by J. Micheals of Oconomowoc, Wisconsion.

45-11546 / N51JW - This aircraft had a number of owners after being purchased as surplus and had a history of flying in the Philippines between 1969 to 1971 as PI-C1046 & RP-C1046. John Wright of San Francisco, California purchased the aircraft and had Aero Sport overhaul the aircraft in 1973. It later crashed in 1982 and was a complete loss.

45-11553 / N51T - the early years saw this aircraft raced as #6 and #15. It was later sold and was rebuilt at Aero Sport in 1970. After an accident it was rebuilt with parts from two other Mustangs and flew and raced as "Miss Fit" #553 / N5415V. The aircraft is now registered as N38JC and owned by Avirex Inc., New York.

45-11558 / N6175C - in 1972 this aircraft was rebuilt from parts of two other dismantled P-51 airframes. The aircraft is currently owned by Joe Kasparoff, of Van Nuys, California, who races it as #39 "The Healer."

P-51D SN44-73206 / N3751D was a Cavalier conversion and one of the many colorful P-51's that was rebuilt at Aero-Sport during the 1970's. Shown carrying F-AZAG markings while operating in Tahiti, after Aero Sport overhauled it. After being overhaul, Les Howard put one of his colorful paint shemes on it. The authentic military style paint schemes didn't catch on in the Warbird community until the early 1980's. The aircraft is currently flying as "Hurry Home Honey" SN44-13586.
(Top photo by Gerald Liang - bottom photo by Philip Wallick)

Very few P-51 Mustangs carry the dual mirrors located on the windscreen frame, as shown here on Elmer Wards "MAN O WAR," which was first rebuilt at Aero Sport. Elmer Ward would later buy Aero Sport and start Pioneer Aero, which still operates as the world's largest Mustang parts warehouse in the world.

P-51D, SN44-63576 / N51DH is currently operated by the Evergreen Heritage Collection, based in Oregon. The aircraft is shown on final approach to runway 21 at Chino airport. Note the pilot wearing just a headset and not a helmet.

P-51D SN44-74756 / N69QF sitting outside the Aero Sport hangar in 1973. This aircraft had one of the most colorful paint schemes worn on a Mustang. Ken Burnstine raced "Miss Susi Q" as #33 and was killed in an accident after the 1976 Mojave National Air Race, Mojave, California. The aircraft was a complete loss. (Photo by Gerald Liang)

After P-51D, SN44-73029 departed U.S. military service it flew with the Nicaraguian Air Force as GN-122. When it entered civilian life, it was raced as #51 and later flew with the Confederate Air Force as N51JB. It currently flies as "Bald Eagle." (Photo by Emil Strasser)

P-51D, SN44-84753 / N251BP is operated by the Palm Springs Air Museum and is part of Bob Pond's collection. The aircraft was also operated as N5436V / N51TC / N51BE.

P-51D, SN45-11540 / N5162V shown parked at the Van Nuys airport in 1962 wearing a very colorful paint scheme.
(Photo by Emil Strasser)

Another photo of P-51D, SN45-11540 / N5162V shown during a complete rebuild at Aero Sport in 1978. The aircraft currently flies as N151W "Queen B."
(Photo by Gerald Liang)

After a complete overhaul, P-51D SN44-73832 / N117E departed the Aero Sport hangar and was probably one of the strangest looking Mustangs painted overall light grey with multi-colored flowers on its tail. The aircraft saw only two owners between 1963 and 1992 when it was lost in an accident in Nevada. Look hard and you will see four other Mustangs in this photo, three in the hangar and another outside, behind N117E. (Photo by Gerald Liang)

Aero Sport was the first West Cost based operation specializing in Warbird aircraft restoration and maintenance. Cavalier Mustang, was based in Florida and had established itself on the East Coast with the conversion of more than 60 P-51's for third world militaries and conversion for civilian use. Aero Sport and Cavalier, both played an important role in putting ex-military aircraft into civilians hands and creating the warbird movement.

P-51D, SN44-73832 / N117E shown parked on the ramp outside the Aero Sport hanger. (Photographer unknown)

P-51D, SN44-73149 / N6240T raced as "Candy Man" #7 and now flies as SN46-3221 / G-BTCD, "Moose," with The Figther Collection, Duxford, England. (Photo by Gerald Liang)

Jerry Morrison owned this Model "A" Ford that was stuffed in the corner of Hangar #2, with one of the North American T-6 / Harvards waiting rebuilt at Aero Sport. It was not uncommon to have a number of aircraft in storage at Aero Sport parked at the West end of the Hangar #2, while the East end was where all the work was performed.
(Photo by Philip Wallick)

A Douglas A-26C when it first arrived at Aero Sport. There is some question as to the aircraft's real identity and the serial number is unknown. The A-26 is well known for being very fast for its weight and size and can carry a great deal of cargo. It is possible that the aircraft was used as a drug runner in the late 1960's and early 1970's. (Photo by Gerald Liang)

Grumman F8F-2 Bearcat BU122619 / N7958C operated with the Confederate Air Force for a number of years. It was purchased by Harold Beal, who had Aero Sport perform annual inspections. The aircraft operated as N700F, is currently registered as N14WB and displayed as U.S. Navy BU122619 at the E.A.A. Museum in Oshkosh, Wisconsin. (Photo by Gerald Liang)

Grumman F8F-2 Bearcat, BU122629 / N777L was rebuilt and had airframe aerodynamic improvements fabricated at Aero Sport. The aircraft is shown during its first engine start and for its first flight after having a Pratt & Whitney R3350 installed at Aero Sport. Lyle Shelton flew "Rare Bear" #70 and later #77 into the aviation history books, setting records and winning races. The aircraft still flies today as one of the most effective air racing aircraft. (Photo by Bob Nightingale)

TBM INCORPORATED

TBM INC. was a Tulare, California based aerial fire fighting Tanker Company under contract with the State of California and the U.S. Forest Service to provide aerial tanker support in fighting forest fires in Southern California. The company operated quite a few different types of aircraft to include Grumman F7F Tigercats, Boeing B-17 Flying Fortress's and Grumman TBM Avenger's as Aerial Fire Fighting Tankers. These aircraft operated from a tanker base located on the south side of the Chino Airport during the 1960's and 70's.

Which air tanker company won the State contract would dictate where their tankers were based during any given fire session. TBM, Inc. always had aircraft based at Chino during the time the facility was operational.

These ex-military aircraft were very effective weapons, dropping chemicals instead of bombs on forest fires in Southern California. The aircraft were operated for more than 20 years before their retirement. The State of California and the U.S. Forest Service wanted newer and bigger multi-engine aircraft, while the operators realized the value of World War II aircraft to collectors. Most of the WWII type aircraft tankers were sold to private owners or traded for more modern aircraft that could be used as air tankers.

Though most have been retired, there are a few operators still flying vintage bombers as Fire Fighters. Air Spray of Canada still operates a large fleet of Douglas B-26 Invaders and Hawkins & Powers in the United States still operate a fleet of PB4Y2 Privateer's.

Over the years, many different types of ex-World War II aircraft were utilized as fire fighting tankers to fight forest fires throughout the United States and Canada. Here is a list of most of the aircraft types used; PT-17 Stearman, Lodestar, PBY Catalina, PB4Y2 Privateer, TBM Avenger, AF-2 Guardian, F7F Tigercat, B-26 Invader, B-25 Mitchell and B-17 Flying Fortress.

Tanker #E61 SN443-38635 / N3702G suffered a main landing gear failure after returning from a fire. Aero Sport repaired the aircraft and it was returned to fighting fires until it's retirement in 1979. The aircraft is currently displayed as SN43-8635 "Virgin's Delight" at the Castle Air Museum, Modesto, California. (Photo by Bob Nightingale)

Tanker #E61 SN443-38635 / N3702G shown sitting alert at the Chino Air Tanker Base in July 1969. (Photo by Gerald Liang)

Another one of TBM's B-17's that sat alert at the Air Attack Base at Chino during the 1970's was Tanker #E78, which also flew as #E68 SN44-83546 / N3703G. After Tanker #E78's retirement, it was purchased by David Tallichet. The aircraft was flown in the movie "Memphis Belle" and is currently on display at the March Air Force Base Museum, California.

Tanker #64, BU80425 / N7235C a Grumman F7F "Tigercat" was operated by TBM Incorporated for over ten years. When retired from its fire fighting role, the aircraft was sold to David Tallichet, who put the aircraft on display at the Combat Air Museum in Topeka, Kansas and later sold the aircraft to the Fighter Collection, Duxford, England.

TBM Incorporated, operated Tanker #58, BU91110 /N6827C, a Grumman TBM, as an aerial fire fighting tanker for many years. Shown sitting alert, awaiting it's turn to drop fire retardent on a Southern California forest fire. Afterwards, Tanker #58 was retired and sold to The Old Flying Company at Duxford, England and later sold to the Alpine Fighter Collection in New Zealand. The aircraft is currently operated as ZK-TBM by Alpine Deer Group, in New Zealand.

RUTH JOHNSON'S FLIGHT SCHOOL

Ruth Johnson became a highly experienced pilot flying as a WASP during World War II. After the war, she continued her flying career by operating a twin Beech under a government contract at Chino Airport. She later starting a flying service and basic flight school during the mid-1970's. During this period, she taught students in a Piper J-3 Cub. Ruth Johnsons' Flying Service operated from hangar #3 next to Aero Sport until the early 1980's. Ruth also owned a North American AT-6 which she raced at the Cleveland National Air Races.

FLOYD WARDLOW AVIATION

Floyd Wardlow had an interesting aviation background in that when he was a child, he lost one arm in an accident. Later in life, he mastered flying and became a ferry pilot during World War II. Floyd and Cliff Panard operated Hawthorne Aircraft in Lucern Valley where they rebuilt Stearman aircraft. Most of their aircraft were purchased from war assets and surplus parts from military auctions around the country.

When the operation moved to Chino, Hawthorne Aircraft had around twenty or thirty Stearmans, most of which were incomplete aircraft. The aircraft were stored in an old chicken coop on the airport and the operation became known as the "Chicken Coop Stearmans." The business sold aircraft as complete flyable aircraft or as projects that the customer finished.

When Cliff died in 1970, Floyd and son Mark, started Wardlow Aviation which they operated until 1995. Their business was located across from Flos' Restaurant, in one of the old CAL-AERO Academy support buildings where they sold a variety of aircraft parts. They specialized in Boeing, Stearman and North American T-6 / SNJ "Texan," but they also carried a lot of different warbird odds and ends like tires, propellers, engines, airframe parts and fuel tanks and prewar antique aircraft parts.

U.S FOREST SERVICE

The U.S. Forest Service operated two Beech T-34's from Hanger #3. These aircraft where flown as forward air controllers to support aerial fire fighting tankers that also operated out of Chino. This support mission only lasted until the firefighting tanker base moved south to Hemet-Ryan Airport. The T-34's were also used to support liaison duties for the state. Mostly using the aircraft for joy rides and sightseeing flights for dignitaries around Southern California. The operation only lasted a few years and was moved to Sacramento Municipal Airport in Northern California, closer to the state capital.

FRANK TAYLOR

Frank Taylor played a role in the early part of the Warbird movement at Chino. Frank was one of the first Warbird owners that got involved in Air Racing at Chino. Frank Taylor and Bill Destefani purchased North American P-51D " SN44-74996 / N5410V and had numerous airframe modifications done before it was raced at Reno as #4 "Dago Red." Frank also raced cars in the form of dragsters and for a short period, housed the operation along with the Mustang in Hangar #1 next to Aero Sport. Johnny Maloney, of The Air Museum, was part of the "Dago Red" air racing crew when it was first raced at Reno.

Frank Taylor and Bill Destefani owned and raced #4 "Dago Red" P-51D SN44-74996 / N5410V. The airframe modifications where done at Frank Taylor's facility at Chino during 1981. "Dago Red" would change owners a number of times, but the aircraft has still remained an effective air racer at Reno. (Photo by Gerald Liang)

fighter imports

Leroy Penhall Aviation, also known as Fighter Imports Inc., was the founder of the Jet Warbird movement in the United States. Penhall ran a first class operation and was the first to locate on the eastern part of the Chino Airport in a new hangar facility.

Leroy Penhall had made arrangements to import 40 ex-RCAF Canadair CL-30's or T-33 and 30 CL-13 or F-86 Sabres. Not all of the aircraft made it to Chino, but fifteen of the T-33's and two F-86's were rebuilt at Chino. One of the T-33's was purchased by Frank Sanders and flown at many airshows around the country as the "Red Knight." Yet another of the other CL-30 airframes was used for the ill-fated Boeing twin engine T-33 "Skyfox," purposed as an inexpensive light attack aircraft for third world countries. Bob Hoover flew one of the CL-13's for airshow demonstrations. One of the CL-13's was later purchased by Jim Robinson/ Combat Jets Museum and flew as "HUFF." The aircraft is currently at the E.A.A. Museum. One of the Cl-13's was sold and operated with Flight Systems Inc, (later Tracor Flight Sytems) at Mojave Airport, California under government contracts for aerial gunnery target towing.

During the early years, Leroy Penhall raced a highly polished North American P-51D, SN44-74502 / N6321T as #33 and later rebuilt the aircraft as N70QF for Ken Burnstein who raced it as "Foxy Lady" #34. The aircraft then was owned by John Crocker as N51VC and raced as "Sumthin Else" #6. After an accident, the airframe was put into storage until its new owner had Square One rebuild it into a TF-51D. The new aircraft now flies as N1451D "Saturday Night Fever." The other Mustang that was raced under the Penhall's supervision was a yellow Mustang, #81 SN44-74012 / N6519D.

Another important contribution that Leroy Penhall made to the warbird scene was hiring a number of individuals that would play an important role in the Warbird movement at Chino. Bruce Glossling worked as Leroy Penhall's shop manger for a number of years. Bruce would later open Unlimited Aircraft Limited in 1973. Unlimited was one of the first companies to acquire ex-military jet fighters for civilain use in the warbird community. Penhall also hired and mentored Steve Hinton and Jim Maloney giving them a sound aviation foundation in 1971. Leroy Penhall would check both Steve and Jim out in the P-51 and later in the T-33.

Chino and the Wabird community suffered a major loss when Leroy Penhall was killed in 1975, while the Beech Duke he was flying crashed while returning from a skiing trip in Colorado. Figther Imports was closed soon after.

One of the P-51D Mustangs owned and operated by Leroy Penhall was 44-74012 / N6519D race #81, shown parked in front of "Fighter Imports" hangar at Chino. The aircraft crashed while landing in 1974 and went into storage until 1976, when a restoration project was started, but not finished. The aircraft was sold and the restoration was finally completed in 1987. All records still show the aircraft registered as N6519D. (All photos these pages by Gerald Liang)

Above, CL-30 RCAF 21375 / N12430 was operated as a test bed by Devonshire Fighters - SMP in the 1970's. Opposite page - The same aircraft was sold to the Harrah Collection Unlimited, Portland, Oregon in 1985, where it operated as H33HW. Leroy Penhall imported a large number of CL-30's from Canada for conversion into the civilian aviation community.   Parked outside the Chino facilty is a line up of four CL-30's N12414 / N12413 / N12415 / N12416 at Figther Imports during 1972.

The Skyfox Corporation of Van Nuys, California first purposed the concept of the "Skyfox," a low budget coin-fighter converted from CL-30 RCAF 21160 / N221SF. The aircraft was powered by two turbofan engines and was to be sold to small western militaries as a low budget coin fighter. Boeing Aircraft Company later sponsored the project and canceled it after it didn't catch on.

Above, Bob Hoover taxiing CL-13A / N8687D at the 1972 Reno National Air Races prior to a flight demonstration. The aircraft is one of the two Sabres that were rebuilt at the Penhall Shops. (Photo by Philip Wallick)

CL-13A, RCAF 23285 / N8686D was painted a very nice white with red when operated by Figther Imports. The aircraft was later sold to Flight Systems and later to Tracor Flight Systems of Mojave, California. The aircraft flew as N87FS and N92FS and was used to tow aerial gunnery targets for the U.S. Air Force until the early 1990's. (Bottom Photo by Gerald Liang)

CL-13A RCAF 23314 / N8687D after being sold to Flight Systems and later Jim Robinson / Combat Jets Flying Museum, Houston, Texas. The aircraft is currently displayed at the E.A.A. Museum in Oshkosh, Wisconsin.

Following is a listing of most of the aircraft that passed through the Penhall shops at Chino: 12 - CL-30's and 2 - CL-13 and 2 - North American P-51's.

Lockheed T-33A / TV-2 / B SN51-4033 / N6633D, currently owned by Shooting Star Productions, Reno, Nevada.

Lockheed T-33A SN56-3667 / N51SR, currently owned by David Tallichet, aircraft is still located on Chino airport.

Canadair CL-30 MK3 RCAF 21160 / N221SF, was converted into the Boeing "Skyfox."

Canadair CL-30 MK3 RCAF 21200 / N12420 sold as salvage 1976.

Canadair CL-30 MK3 RCAF 21221 / N12424

Canadair CL-30 MK3 RCAF 21265 / N12418 crashed in 1994.

Canadair CL-30 MK3 RCAF 21273 / N12413 later became N233RK, the ill-fated "Red Knight."

Canadair CL-30 MK3 RCAF 21295 / N12419 and later N72JR / currently displayed at E.A.A. Museum in Oshkosh, Wisconsin.

Canadair CL-30 MK3 RCAF 21341 / N12422

Canadair CL-30 MK3 RCAF 21369 / N12416 sold to Boeing Equipment Holding Company, Seattle, Washington.

Canadair CL-30 MK3 RCAF 21375 / N12430 operated with SMP - Devonshire Figthers later sold to Harrah Corp N33HW.

Canadair CL-30 MK3 RCAF 21440 / N12417, currently painted as a "Thunderbird."

Canadair CL-13A MK5 RCAF23285 / N8686D retired Flight Systems target tow.

Canadair CL-13A MK5 RCAF23314 / N8687D ex-Jim Robinson, Combat Jets Flying Museum, "Huff" is currently displayed at the E.A.A. Museum, Oshkosh, Wisconsin.

North American P-51D SN44-74012 / N6519D

North American P-51D SN44-74502 / N6321T parts from this aircraft and SN44-74446 / N1451D currently make up TF-51 / N1451D "Saturday Night Fever."

UNLIMITED AIRCRAFT LIMITED

After the close of Fighter Imports, Bruce Gossling opened Unlimited Aircraft Limited and operated out of Hanger #4.   Prior to this, Bruce had worked for Leroy Penhall as the shop general manager, primarily in charge of the P-51 reconstruction.   Over the next 14 years, more than 35 P-51 Mustangs would receive major maintenance or would be completely rebuilt at Unlimited.   There were also a number of Mustangs modified for air racing at Unlimited in the early 1980's.

Some P-51s were brought back from the Dominican Republic and rebuilt at Unlimited and were later sold to private owners.   Other warbird type aircraft were also worked on at Unimited.

Unlimited was partly responsibile for starting the modern jet movement in the United States.   In 1986, Bruce Gossling went to China to acquire the first batch of Mig-15's to be imported into the U.S.   After the aircraft arrived, Unlimited Aircraft Limited modernized them and brought them up to F.A.A. standards.   Later, more people imported Mig-15's, Mig-17's, a single Chinese Mig-19 and Mig-21's to the U.S. from Poland and the Former Soviet Union.   These aircraft are currently displayed at museums or are owned and flown by private owners.

Currently, there a variety of different types of jet aircraft being flown by private owners.   These include: Aero L-29's and L-39's, BAC Provost and Lightnings, Cessna A-37 Dragonfly, DeHavilland Vampire, Douglas TA-4J Skyhawk, Fouga Magisters, Folland Gnat, Grumman F9F Panther, Hawker Hunters, Hispano HA-200's, Lockheed F-104 Starfighters, Lockheed T-33 Shootingstars, McDonald-Douglas F-4C Phantom, Mikoyan-Gurevich Mig-15 and Mig-17's, North American F-86 Sabres and F-100 Super Sabres, Northrop T-38 Talons, F-5 Freedom Fighters and others.

The North American F-86A RCAF 23314 / N8687D and F-2 Chinese Navy 1411 - Mig-15 / N15MG once flew as rivals during the Korean Conflict. They are now sought after by private collectors and owners. These two aircraft were owned by Jim Robinson, Combat Jets Flying Museum, Houston, Texas. Both aircraft are now part of the E.A.A. Museum, Oshkosh, Wisconsin.

One of the first Mig's to enter the airshow aerial demonstration arena was Paul Entrekin's Chinese F-2 / Mig-15, N90601, which was imported by Bruce Gossling and rebuilt at Unlimited.

One of the first Mig-15's imported from China to the United States sits covered in the Unlimted Aircraft Limited's storage yard near hangar #3 at Chino.

One of the Lim-5 / Mig-17's brought into the U.S. from China by Unlimited Aircraft Limited. No record of serial numbers or registration could be found.
(Photo by Gerald Liang)

Paul Entrekin's Chinese F-2 / Mig-15, first registered as N90601, then N15PE. It was also one of the first Mig's certified by the F.A.A. after being imported into the U.S.
(Photo by Gerald Liang)

Unlimted Aircraft also rebuilt a number of Northrop T-38's and F-5's. One of the first T-38A "Talons" in civilian operations was Chuck Thorton's N5228. The aircraft was built from a number of salavaged airframes.
(Photo by Gerald Liang)

One of the first high performance ex-military fighters in civilian hands was this Douglas A-4A "Skyhawk" BU14219 - N444AV. The aircraft was rebuilt from two different airframes and numerous parts. Originally built for Pascal Mahvi / Aeronautical Test Vehicles in 1984 and sold to Guy Neeley / Advanced Aero Enterprises, Inc. in 1987. The aircraft crashed near Mojave while making a television commercial.
(Photo by Gerald Liang)

The second Douglas "Skyhawk" that rolled out of the Unimited Shops was A-4B BU142112 / N3E rebuilt for warbird jet collector, Jim Robinson's "Combat Jets Flying Museum," based in Houston Texas. The Combat Jet Museum was disbanded and aircraft were donated to and are on display at the E.A.A. Museum.

Unlimited Aircraft Limited operated from hangar #3 during which time these aircraft emerged from the their hangar :

Aero Vodochy L-29 Delfin
Boeing Stearman, N--52V
Canadair CF-104D, RCAF12637 &
        RCAF12633 / N104JR
Canadair CL-13 (F-86), RCAF 23314 / N8687D
Chinese F-2 (Mig-15, 5 total airframes)
        N90589 / N51MG / N15MG
Curtiss P-40E AK899 / N9837A
Douglas A-26C, N26VC
Douglas A-4A&B Skyhawk, BU14219 / N444AV
        & BU142112 / N3E
Vought F4U, NX33693
Grumman F8F Bearcat, BU121752 / N800H
        & N700F
Hispano Ha 200, Spanish Jet Trainer
Lockheed T-33'S, N12418, N33HW, N33VC
        N231, N33EL, N155X, N33DH, N84TB
        N12418, N12414
Mikoyan-Gurevich MIG-15 / N15PE
Mikoyan-Gurevich Mig-17
Mikoyan-Gurevich Mig-19 recovered from
        China
Mikoyan-Gurevich Mig-21 recovered from
        Poland
North American F-86, N86Z & N86CD
North American P-51D's,
        N6344T, N4CL, N6175, N7345V
        N5415V, N3580, N1451D, N51JB
        N51MR, N79111, N51RR, N51EA
        N51DM
North American TF-51, N1451D
North American T-28D, N138NA
Northrop F-5A, N586PC
Northrop RF-5A, N685TC
Northrop T-38A, N5228 & N638TC
Republic F-84F, N84JW
Seata, N86Y
Supermarine Spitfire, NX749DP

Opposite page: Another aircraft rebuilt at Unlimited for Combat Jets Flying Museum was this Canadair TF-104D / RCAF 12633 / N104JR. The aircraft has since been sold and is flown on a regular basis by its new owner.

SANDERS AVIATION

Frank Sanders was born into an aviation family. His father flew Boeing B-17's during World War II. Frank Sanders began flying at a young age and later became famous for his airshow performances and knowledge as an aeronautical equipment designer. Frank's first aircraft was a North American T-6 Texan and later he purchased a Beech T-34. Wanting more performance, he moved on to aircraft like the Hawker Sea Fury, North American P-51 Mustang and finally the Canadair CL-33. The "Red Knight" became the aircraft that Frank made most of his airshow performances in.

Sanders Aircraft was established in 1970 at the Long Beach Airport and later moved to Chino. The Sanders family became known for their specialty work with Hawker Sea Furies and airshow demonstrations.

When the Sanders family moved to the Chino airport they occupied the same hangar that Leroy Penhall had built, just north of the control tower. The first Hawker Sea Fury that Frank Sanders purchased was in 1968, a FB MK.11 TG114 / N232J and the aircraft was raced the following year at Reno. Then came the workhorse of the Sanders family of Sea Furies, "924" a stock dual-flight controlled T MK.20 VX300 / N924G, with its counter-clockwise turning 3,270 cubic inch Bristol Centaurus engine. The aircraft has flown at airshows and has been used as a flying testbed. The aircraft has dual flight controls and has been used by many military test pilot programs for check out in a high performance propeller driven aircraft.

"The Red Knight" Canadair CT-33 flown by Frank Sanders became famous for its aerial flight demonstrations at Airshows around the country. (Photo by Gerald Liang)

Two of the Sanders Hawker Sea Fury's - N924G - "924" as she is called and the newest Sea Fury in the Sanders stable N19SF - "Argonaught" - race #19 sitting on the ramp near their hangar at Chino.

Both of Frank's sons Brian and Dennis, were also checked out in "924" at the early age of seventeen. The next Sea Fury was the highly modified racer "Dreadnaught" T MK.20 - VZ368 / N20SF, race #8, which was powered by a 4,000 horse power Pratt & Whitney 4360 engine and was pulled through the skies by a AD-5 Skyraider propeller.

Frank Sanders designed a smoke generator system that airshow performers could use on the wingtips of their aircraft. The system would produce two ribbons of smoke while the aircraft flew through the skies. The system is used by a number of countries on demostration aircraft around the world, to include Lockheed F16 Fighting Falcon's, Boeing F/A-18 Hornet's and F-15 Eagle's, Northrop F-5's and Mirage F-1's and 2000's and many other militaries aircraft. The system was further developed into a low cost air-to-air combat training concept, which used a laser signal instead of bullets to shoot down an opponent during aerial combat training missions. The system was called the "Acewinder." The concept was simple, once the laser locked onto the target, a smoke trail would identify that he had been shot down. It was a very low budget training tool for any militaries air force, but it never caught on.

Tragedy struck, when Frank Sanders was killed while flying the "Red Knight" in New Mexico in 1990. This was a great loss to the family and to the Warbird community. Since then, Brain and Dennis have continued with the family business and continued performing at airshows in the aircraft that made the "Sanders" name synonymous with the Hawker Sea Fury.

Sander's Aviation, and the Sanders family played an important role in the Warbird scene at Chino and though they have moved their operation to Northern California, they still continue to build Hawker Sea Furies and other Warbirds at their new location.

Over the years, a number of aircraft emerged from the Sanders hangar. Following is a listing of them:

Beech T-34A, N34Z
Boomerang, VH-BOM / NX4234K
    NX32CS
Commonwealth C-27 / F-86, (4)
A94-909, A94-914, A94-916, A94-354
Curtiss P-40N, 44-7619 / N222SU
Fiat G-59-4B, NX59B "CINO BELLA"
Harvard, MKII AJ731 / N9789Z
Hawker Sea Fury, N20SF
Hawker Sea Fury, N19SF, race #19
Hawker Sea Fury, N924G
Hawker Sea Fury, NX20SF
"DREADNAUGHT"
Hawker Sea Fury, WJ288
Naval Aircraft Factory N3N-3
North American P-51D
    SN44-74996 / N5410V
North American T-28B
    BU137799 / N28YF
North American AT-11 (2)
Sea Venom, WZ944 / N7022H
Vodochy L-29 Delfin, CN 591699 /N179EP
Yakovlev LET-C.11 / N2124X

One of the different types of aircraft that was built at the Sanders facility, was this Fiat G-59-4B / NX59B "CINO BELLA," shown during a local flight near Chino. (Photo by Gerald Liang)

Commonwealth Aircraft Corporation CA-19 A26-206 / NX4234K / VH-BOM, is shown under construction in the Sanders hangar when they were located at Chino. This aircraft was later sold and shipped to Australia.

Sanders Aviation modified two T-6 airframes into flyable CAC Boomerangs. Initially, Dennis Sanders and Dale Clark were partners on CA-13 A46-139 / N32CS "Phooey" shown during a flight near Chino. (Photo by Jerry Wilkins)

Hawker Sea Fury 232 is shown taxiing out for one of the many aerial demonstrations that Frank Sanders performed in this aircraft at airshows around the country.

Hawker Sea Fury "924" flying over the San Gabriel Mountain range near Chino. Aircraft "924" has been the workhorse for the Sanders family for more than twenty years, flying as a pilot trainer, flying test bed, airshow demonstrator and as an Unlimited Air Racer. "924" still flies with the standard five bladed propeller and Bristol Centaurus engine. Below, not normally utilized, "924" is shown with wing tanks while taxiing at an airshow in Northern California.

Both U.S. Navy and Air Force test pilot schools utilized "924" for training test pilots in high performance propeller driven aircraft. It was also used for flight testing the Sanders designed "Smoke Generator" and "Ace Winder" air-to-air combat training missile systems. Above, the "Acewinder" missile is shown attached to "924" on an underwing pylon during flight testing.

PACIFIC FIGHTERS

Pacific Fighters was another warbird company that operated at Chino. It was owned and managed by John Muszala between 1988 and 1996. John started his aviation career at The Air Museum Planes of Fame, where he worked for ten years. During this period, John was also part of the RB-51"RED BARON" racing team. John would later race a P-51 and Hawker Sea Fury at the Reno National Air Races. After working for the Air Museum, John went to work for Bruce Gossling at Unlimited Aircraft Limited as the shop manager. John's break came when he received financial backing from a business associate at the airport and "Pacific Fighters" was born. Pacific Fighters operated for a number of years and built quite a few different types of aircraft to award winning standards. Then, in 1996, the decision was made to move the operation to Twin Falls, Idaho where Pacific Fighters now operates and continues to build warbirds.

Don Hanna pulls his AD-4, BU126959 / N2088V Skyraider almost vertical during a wing over maneuver while flying near Chino.

A Douglas B-26 Invader and AD-4 Skyraider, North American P-51 Mustang and T-6 Texan can be seen in "Pacific Fighters" hangar while at Chino. The hangar is now utilized by "Aero Trader."

Pacific Fighters imported a number of ex-French Air Force Douglas AD-4 Skyraiders. Shown is an AD-4W, radar and electronic version, BU126867 / N4277N, owned and operated by Ekickson Air Crane of Oregon.

Grumman F7F-3, BU80483 / N6178C Tigercat is owned by Dick Breta. The aircraft once flew for Cal-Nat Airways as a Fire Fighting Tanker, #E43 while fighting forest fires in California.

Grumman C-1, BU146048 / N7171M "COD" was maintained and operated by Pacific Fighters. The aircraft was owned by Dick Breta, Don Hanna and John Muszala, who all flew it to a number of airshows in the western part of the United States. The aircraft is currently owned by Bob Pond and is displayed and operated by the "The Palm Springs Air Museum" Palm Springs, California.

This is one of only six examples built by Douglas of the XA2D-1, BU125485 "Skyshark." The aircraft was powered by twin T-40 turbo-jet engines and had twin contra-rotating propellers. The aircraft was orginally acquired from salvage by David Tallichet and later sold to John Muszala, who still has it in storage. It is 80% complete and with a lot of time and money could be restored back to flyable condition.

Over the years, Pacific Fighters restored many different types of aircraft while operating at Chino, the following is a list of most of them:

    Canadian Harvard IV
    Douglas AD-4 Skyraiders,
        BU126867 / N4277N
        BU127931 / N4277L
        BU126912 / N4277P
        BU135152 / N65164
        BU126959 / N2088V
        BU126935 / N20088G
    Grumman F7F-3 Tigercat
        BU80483 / N6178C
    Grumman C-1A BU146048 / N7171M
    Hawker Sea Fury T MK.20S
        VX302 / N51SF
    Hawker Sea Fury MK11
        WM-483 / N42SF
    North American AT6G / N2055G
    North American P-51D Mustang
        SN44-84634 / N51ES
        "Big Beautiful Doll"
    North American P-51D
        44-73350 / N33FF
    North American FJ-3 Fury (Static only)
    Vought F4U-5 Corsair
        BU124486 / N49068

Dick Breta, had Pacific Fighters build one of the most impressive Hawker MK11 WM-483 / N42SF Sea Fury restoration projects flying today.

Above and opposite page: "Big Beautiful Doll," owned by Ed Shipley, was one of the first major restoration projects that Pacific Fighters completed. There was an extreme amount of time spent on detailing items to make them as authentic as possible and then some. All to make it one of the most pristine P-51's flying today.

472218
WZ I
Big Beautiful Doll
Col J.D. Landers

YESTERDAY'S AIR FORCE / MARC

David Tallichet owned and operated one of the largest collections of ex-military aircraft in the world. Known as Yesterday's Air Force, only operated for a short period of time at Chino airport. Though headquartered at Chino, aircraft were scattered all over the United States. At one time it is estimated that there were more than two hundred aircraft in the collection. Most of the collection of aircraft has since been dissolved with the aircraft either being sold or on loan to other aviation museums around the country. Tallichet operation is still based at Chino airport with a limited number of aircraft, most in storage, but a few are still used in T.V. commercials and movies. The exact number of aircraft left in the collection is not known and there are still a number of aircraft for sale.

David Tallichet started his aviation career in the Army Air Corps pilot training program in June of 1944. After graduation from basic flight school, he went on to twin-engine advanced aircraft training. Upon graduation, he found himself crossing the Atlantic in a B-17 headed for Europe, where he flew with the 100th Bomb Group flying 22 combat missions and completing his European tour flying C-47's in Germany. After returning from Europe, he went into the Air Force Reserve system and flew T-6 Texans and P-51 Mustangs. Soon after David went into the inactive reserve because the military would not let him fly anymore. The other factor that was his professional career was demanding more time and was becoming more important than the military.

It wasn't long after his success in the business world that he was lead back into aviation. The first aircraft David Tallichet purchased was a Stearman, which he still owns. Soon after co-signing a loan for a friend to buy a Mustang and spare parts for $14,000, Tallichet was again flying a Mustang for just the gas and oil. Then when the owner of the Mustang failed to make the payments, Tallichet was notified by the bank that he was now

David Tallichet shown flying his first P-51D SN44-63893 / N3333E near Cal-Aero Field, California. When this aircraft was first acquired, it was painted in pure civilian markings and was later changed to the military colors shown. (Photo from David Tallichet's collection)

The Indian Air Force flew this Consolidated B-24 SN44-44272 for a number of years before its retirement and donation to Yesterdays Air Force, which also flew it for a number of years before selling it to Kermitt Week's, Fantasy of Flight Museum in Florida. (Photo from David Tallichet's collection)

the proud owner of a P-51 Mustang. This was the primary reason that David Tallichet got started collecting aircraft. The aircraft was sent to Aero Sport at Chino for partial rebuild and maintenance.

Bill Ross, another aircraft collector, convinced Tallichet he should start collecting vintage aircraft as an investment. Tallichet then stated collecting aircraft as Yesterdays Air Force, which later changed to the Military Aircraft Restoration Corporation (MARC). They continued buying aircraft and went public on the stock exchange in 1970 and later converted back to a private company in 1980. During these years, aircraft were acquired from around the world. The Indian Government provided the collection a very rare, flyable Consolidated B-24 Liberator, while three Martin B-26's came out of crash sites in British Columbia and 20 different types of aircraft were recovered out of New Guinea. A number of Republic P-47's Thunderbolts came from the Nicaraguan Air Force and a group of ex-Vietnamese Air Force Douglas A-1 Skyraider's come out of Laos and Cambodia after the Vietnam Conflict. Some Mig-15's and Mig-17's were brought into the country and dispersed in aviation muse-

ums around the country. One of the biggest purchases and recoverd projects was getting 24 Hawker Sea Furies out of Iraqi.

Some of the warbirds were traded with the U.S. Air Force for use in museums at air bases around the country. In trade MARC received recently retired but operational C-123 Providers and a few other aircraft. Other aircraft were being restored and became part of the MARC collection.

If you look hard enough you can see the remains of three Bell P-39 Airacobras. These aircraft remains are currently located at the Chino facility storage yard.

One of the more unique aircraft acquired for Yesterday's Air Force was this Hawker Hurricane MK XII RCAF 5390 from Canada, which was rebuilt by RRS Aviation in Texas.
(Photo from David Tallichet's collection)

The Supermarine Spitfire FR MK XIX PM627 was completely restored in England by Craig Charleston. The aircraft was dismantled, sent to Chino, reassemblied and flight tested. It didn't remain in the collection very long as it was sold to a private party in Colorado. Unfortunately only months after, aircraft and owner were lost in a accident.

Over the years, many changes occurred with the Tallichet collection, it was continually being moved and aircraft were bought and sold on a regular basis. Some aircraft were used in big screen movies like "Catch22" and "Memphis Belle," while others were used in television commercials. Then, in the late 1980's, the company was forced to sell a number of the aircraft and an additional sixty or more aircraft were sold in 1995. In both cases, the reason was to pay off the high operating costs and taxes that had accumulated over the years.

Currently MARC has a number of aircraft loaned to aviation museums around the country. There are also a number of other interesting aircraft being restored back into flyable condition at locations around the country. These aircraft consist of a Douglas A-20 Havoc, which is close to flying, a Martin B-26 Marauder, 2-Douglas SBD Dauntless's, a Curtiss P-40, and 2-Bell P-39 Airacobras. MARC is still looking to recover more aircraft as they are found around the world.

MARC sold this Fairchild C-123K to Aero Union and later TBM Inc. operated it as the only one of its type flown as a firefighting tanker.

There are still quite a few aircraft maintained in storage or that are in flyable condition at locations around the country. These include a L-5, Stearman, T-6, B-17, 2-B-25's, A-26 and a P-51. Though most of the collection has come and gone, the importance that David Tallichet played in the Warbird community has been and still remains, an important part of Chino.

One of three Lockheed P-38L, SN44-27231 / N79123, was restored as Maj Dick Bong's aircraft "Margie," America's highest scoring Ace in World War II. The aircraft sat on display at March Air force Base Air Museum, California for a period of time before it was sold to the newest aviation collection being put together in Seattle, Washington. Another P-38L SN44-53015 is displayed on a pedestal at McGuire Air Force Base, New Jersey as Maj McGuire's P-38 "Pudgy" and the third P-38L, SN44-53186 is now owned and flown by the Heritage Collection in Oregon.

This was the first of three Lockheed P-38L Lightnings, that passed through Tallichet's operation. SN44-53015 / N9957F is shown parked between the Aero Sport hanger and #3 hangar where Tallichet started his operation.

There were only 238 - O-47's built for the U.S. Army Air Corps. The current location of this aircraft is not known. A North American O-47A "Owl," serial number is unknown, was restored by the Military Aircraft Restoration Corporation.

North American P-51D SN44-74978 / N6169U, shown parked on the ramp near Aero Sport. This aircraft was sold as N74978 after it was rebuilt and in 1964. The aircraft was lost in a hangar fire at Shafter Airport, California in 1988.

Stinson L-5 SN42-96468 is still owned by Tallichet / MARC and is currently stored at the Chino facility.

One of three Bristol Blenhiem Boingbroke that was acquired over the years and probably the only one that came close to being restored. The aircraft is parked at Tallichet's facility while in #3 Hangar during the early years at Chino.

Over the years, Tallichet / MARC acquired two Vought Corsairs, both were FG-1D's - BU92106 / N6897, shown here is in storage awaiting restoration & BU92132 / N3466G owned by Butch Schroeder / Midwest Aviation Museum.   The Bell P-63A, Kingcobra SN42-69097 / N52113, is now owned and displayed at The Figther Collection, Duxford, England.   (All photos these two pages by Gerald Liang)

Three Martin B-26 Marauders were recovered sitting in the weeds in Smith River, British Columbia, Canada where the aircraft made forced landings in 1942 and sat ever since. After being recovered, the aircraft were moved to California. (Above photo from David Tallichet collection). B-26 SN40-1464 was completely rebuilt to flyable condition. The aircraft was flown only a few times before being sold to Kermit Weeks. Aero Trader then made numerous maintenance changes and improvements to the aircraft before it was flown back to Kermit Week's Fantasy of Flight museum in Florida.

The back storage yard at Tallichet's Chino facility, houses parts of aircraft which could be made into complete aircraft.

YESTERDAY'S AIR FORCE
AIRCRAFT LISTING

The following is as complete a list as possible of most of the aircraft that were owned at one time or another by David Tallichet, AKA Yesterday's Air Force Museum, AKA Military Aircraft Restoration Group. These are as many of the aircraft types, (quantities, serial number or bureau number and U.S. civilian registration number as I could find.

Bell P-39 Airacobra (15) SN42-8740 / N81575
    SN42-4949 / SN42-18403 /SN42-18408
    SN42-18547 / SN42-18811 / SN42-18814
    SN42-19027 / SN42-19034 / SN42-19991
    SN42-19995 / SN42-20000 / SN42-20007
    SN42-20339 / SN44-2438
Bell P-63 Kingcobra (2)
    SN42-69097 / N52113 / 43-11117
Boeing B-17G Flying Fortress (2)
    SN44-83546 / N3703G
    SN44-83663 / N47780
Boeing B-29A Superfortress (1)
    SN44-61669 / N3299F

Brewster A-34 Bermuda (3)
    MK1 FF860, 2- MK1's Serial # Unknown
Bristol Beaufort (6)
    MKVIII A9-210 / A9-414 / A9-535 /
    A9-559 / A9-637 / RAF DD931
Bristol Blenhiem / Fairchild Bolingbroke (3)
    MK.IV RCAF 9048 /
    MK.IVT RCAF 10073 /
    MK.IVTT RCAF 10076
Consolidated B-24 Liberator (3)
    SN42-40461 / SN42-40557
    SN44-44272 / N94459
Consolidated PBY-5A Catalina (5)
    BU46456 / N4582T, BU46457 / N4582U
    BU46582 / N4583A, BU46595 / N4583B
    BU46624 / N9502C
Curtiss 052 Owl (2)
    SN40-2769 / N61241, SN40-2746
Curtiss P-40 Warhawk / Kittyhawk (9)
    AK899 / N9837A
    FX760 / SN42-104818 / SN42-104959

This Boeing B-17G is still flown on a regular basis and a workhorse in the Tallichet fleet. Tallichet flew B-17's in WWII and says that this would be the last aircraft he would sell.

SN42-104961 / SN42-105710
SN42-105951 / SN42-106101
SN44-7983 / N9950

Curtiss SB2C Helldiver (1)
    BU19075 / N4250Y
DeHaviland Mosquito (1)
    B MK.35 RS709 / N9797
Douglas SBD-4 Dauntless (1)
    BU10715
Douglas A-20 Havoc (2)
    SN43-21627 / N99385 & SN44-20?
Douglas A-24B (1)
    SN42-54654
Douglas A-1H/E AD-4/5/6 Skyraider (5)
    BU127922 / N5469Y
    BU132683 / N39147
    BU135332 / N39148
    BU139606 / N39606
    BU139665 / N39149
Douglas A2D Skyshark (1)
    BU125485
Douglas A-26 Invader (7)
    SN43-22374 / N6843D
    SN43-22602 / N99900Z

SN43-22668 / N99422
SN44-34104 / N99420
SN44-34697 / N4807E
SN44-35888 / N4810E
SN? - N99425
Fairchild C-123 Provider (1)
    N123DT
Folland Gnat / Hindustan HF 24 Adjeet (7)
    MK1 IE1076 / MK1 IE1214
    MK1 IE1222 / Ajeet E276 / Ajeet E296
Ajeet E299 / Ajeet E315
Grumman F6F-5K Hellcat (1)
    BU80141 / N100TF
Grumman F7F-3P Tigercat (1)
    BU80425 / N7235C
Grumman TBM Avenger (3)
    BU53454 / N7030C
    BU53804 / N9710Z
    BU91450 / N7239C
Hawker Hunter MK.50 (1)
    FV34006 / MK.51 35-418
Hawker Sea Fury (21) FBMK.11 - WJ298 / N26SF
    FB.10 / N21SF, FB.10 / N24SF
    FB.10 / N3OSF, FB.10 / N28SF
    FB.10 / N34SF, FB.10 / N35SF
    FB.10 / N36SF, FB.10 / N38SF
    FB.10 / N40SF, FB.10 / N43SF
    FB.10 / N45SF, FB.10 / N46SF
    FB.10 / N48SF, FB.10 / N54SF
    FB.10 / N54SF, FB.10 / N56SF
    FB.10 / N58SF, FB.10 / N62SF
    N63SF / N64SF, FB MK.11 / N54SF
Hawker Hurricane - RCAF 5390
Heinkel HE111 - CASA 2.111 (1)
    EB21-27 / N99230

Junkers Ju52 - CASA 352 (2)
- T.2B-140 / N9021P
- T.2B-142 / N9012N

Lockheed P-38 Lightning (3)
- SN44-27231 / N79123
- SN44-53015 / N9957F
- SN44-53186 / N505MH

Lockheed T-33A Shooting Star (1)
- SN56-3667 / N51SR

Martin B-26 Marauder (4)
- SN40-1459 / N4299K
- SN40-1464 / N2497J
- SN40-1501 / N4299S
- SN41-31856 - ?

Messerschmit Bf109 - Hispano Ha 1112 M1L (1)
- C.45K-75 / N3109G

Mikoyan-Gurevich Mig-17 / LIM-5 (2)
- Never registered

Mikoyan-Gurevich Mig-19 (1)
- Currently displayed at the March AFB Museum

North American AT-6 Texan (3)
- SN44-81857 / N81854
- Other two unknown

North American T-28 Trojan (17)
- SN49-1540 / N99395, SN49-1645 / N99394
- SN51-3513 / N9868A, SN51-3529 / N?
- SN51-3530 / N8523A, SN51-3557 / N85228
- SN51-3565 / N8522Z, SN51-3627 / N8539A
- SN51-3690 / N54612, SN51-7536 / N99393
- SN51-7669 / N9860A, SN51-7730 / N9857A
- SN52-1226 / N8522X, BU137745 / N9038L
- BU137782 / N9039Z, N99412, N99414,

North American B-25 Mitchell (5)
- SN44-29366 / N9115Z
- SN44-30210 / N9455Z
- SN44-30324 / N3161G
- SN44-31032 / N3174G
- SN44-86701 / N7681C

North American P-51 Mustang (4)
- SN44-63893 / N3333E
- SN44-74494 / N6356T
- SN44-74978 / N6169U
- SN44-11807 / N30991

Notrth American O-47A (1)
- SN-Unkown

Republic P-47 Thunderbolt (7)
- SN44-90471 / N47DA

Currently, there are no flyable Douglas A-20 Havocs on the Warbird scene. Tallichet's Military Aircraft Restoration Company has one almost ready to fly and another that is capable of being completely restored to flyable condition with spare parts in his storage yard at Chino.

A very rare Curtiss O-52 "Owl" SN40-2746, is complete and for sale. The aircraft could be restored to flyable condition and is currently stored at Chino.

SN44-49181 / N47DC
SN44-49192 / N47DD
SN45-49167 / N47DB
SN45-49205 / N47DE
SN45-49385 / N47DF
SN45-49509 / N47DD

Supermarine Spitfire (5)
LF MK.IXE / NH23B, N238V
LF MK.XVI RW382 / N382RW
FR MK.XVIII TP298 / N41702
PR MK XIX PM627 / ?
FR MK XVILLE SM845 / G-BUOS

Vought FG-1D Corsair (2)
BU92106 / N6897
BU92132 / N3466G

Westland Lysander (1)
MKIIIA RCAF

MARC imported most of the aircraft that were acquired from overseas countries. Some of the types of odd and rare aircraft imported were; Iraqi Sea Furies, Soviet and Polish Mig-15's, Mig-17's and a Mig-19, Douglas A-1 Skyraiders from the Vietnam Conflict and aircraft from some South Pacific Islands.

Over the years, there were probably more aircraft than those listed, as well as large group of parts and other odds and ends that have passed through the Tallichet Collection / MARC.

There is no question that the Tallichet collection / MARC at one time was the largest civilian owned collection of ex-military aircraft in the world, having more aircraft than some countries air forces operate.

This was the only Mig-19 imported by MARC into the U.S. (Photo by Gerald Laing)

5 STEARMAN OPERATIONS AT CHINO

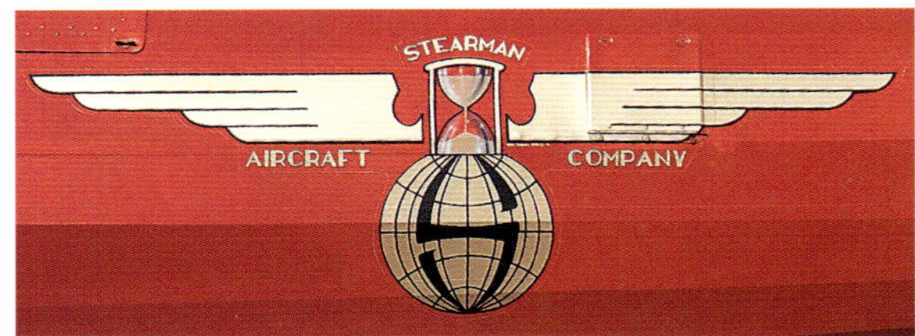

The Stearman Aircraft Company designed the PT-series aircraft in 1934. Stearman later became part of the massive Boeing Aircraft Company, which continued the production of the Stearman. The U.S. Army Air Corps purchased PT-13 and later PT-17 versions of the Stearman as their primary pilot training aircraft. Comparable versions of the PT series where also purchased by the U.S. Navy as the N2S. The U.S. military utilized more than 9,000 Stearmans for training. Its primary mission was to train military pilots and it did so well, it became known as the "Pilot Maker."

The Stearman was the primary aircraft used as a basic flight trainer, to train thousands of military pilots at CAL-AERO Academy and other military and civilian flight schools. The Stearman was also exported to many other countries that utilized it for many years.

After the war, the Stearmans were sold off as military surplus, just like all the fighters, bombers, and other excess military aircraft. Many of these aircraft were bought from the military at salvage yards around the country. Because of it slow maneuverability and flight characteristics, the Stearman was a perfect aircraft for many civilian related applications. Some of these applications

There are approximetely 18 Stearmans stuffed in hangers on the Chino airport.

They come in a variety of colors and carry many different markings. Weather permitting on any Saturday you can see a Stearman in the pattern or operating to and from Chino.

being in agriculture, where the aircraft was modified to drop new seeds and spray crops, commonly known as, "Crop Dusters." Some of the first aerial fire-fighting aircraft were Stearmans, modified with an external tank to drop water on fires. Yes, these aircraft were also showing up at civilian flight schools.

Currently, the F.A.A. shows about 1,000 Stearman airframes on the books, but it is estimated that only about 400 aircraft are registered and flyable around the world. That would leave a lot of airframes and parts sitting in barns, sheds, garages and hangars around the country awaiting their chance to get back in the sky. The largest group of Stearmans can be seen at the National Stearman Gathering, which is held every September at Galesburg, Illinois. This annual get together of Stearman owners brings over 100 aircraft together of every variety you could imagine. Purely stock and original to highly modified, custom aircraft.

The Stearman movement at Chino is an important part of the history of the airport. Not only was the Stearman the primary pilot trainer used at CAL-AERO, but it has continued to be an important part of the Warbird scene. There are more than 15 flyable Stearmans at Chino with more under restoration or rebuild. The following information is a fraction of the history of the Stearman at Chino.

Opposite page - The business end of the aircraft, the engine, Stearman were powered by Pratt & Whitney Wasp and Continental engines, depending on the model of aircraft.

VINTAGE AEROPLANE COMPANY

Tony Farhet founded Vintage Aeroplane Company in 1978. After deciding that he wanted a Stearman, Tony went to Floyd Wardlow, who had a number of aircraft in various stages of rebuild at Chino airport. When Floyd wasn't building the aircraft fast enough for Tony, he took it upon himself to build the aircraft with verbal help from Floyd Wardlow. The project took a little more than a year to finish in 1980. Tony flew this aircraft for seven years before he sold it. His second Stearman was purchased in 1987 in nearby Corona, where it had been stored since the owner purchased it after the war. The aircraft was totally original and had never been used as a crop duster. Many of the parts were new and still in original Boeing wrapping paper. Soon after the aircraft was completely rebuilt and flying, a client purchased it from Tony. The third aircraft was only partially built when it was purchased. During 1970 and 1980 Stearman parts where plentiful because of the partial retirement of Stearmans from the crop dusting community. Tony started purchasing parts and aircraft frames from wherever he could find them.

During the past twenty years, Tony has flown more than 2200 hours in Stearmans and built a total of ten aircraft - each taking about a year to construct. Tony sub-contracted out the engine and wing covering fabric work, but completed the rest of the restoration and construction himself. Tony continues to supply Stearman operators around the world with parts and service, but has discontinued building Stearmans. As important, are the uncounted Stearman projects around the world that "Vintage Aeroplane Company" has supplied parts to complete an aircraft.

There are usually one or two Stearman under construction in one of the hangars on the airfield. Once a crop duster, Stearman SN75-1546 / N51809, when completed, will fly as N450KH and will be better than it was when it rolled off Boeing's production line in 1940. The aircraft has gone through a ten year restoration project and is almost complete. The quality of workmanship can be seen throughout the aircraft. Like most aircraft projects, the total cost of a project is at the owners discretion. In this case, the cockpit alone, cost more than Boeing's original selling price of a complete Stearman in 1940.

STEARMAN FLIGHT CENTER

Hartley Folstad started flying and building Stearman's in the early 1970's. Like most of us he wanted a P-51, but only had enough for a Stearman. This proved to be totally acceptable. Since he acquired his first PT-17, he has logged more than 3,000 hours, has assembled 8 aircraft and currently owns six PT-17's. These are: N450JN / N450SR / N622SR /N621SR / N65263 and one currently under construction. Hartley is also having two T-6 rebuilt at Chino by Dave Hansen.

Carter Tweeters taxis one of Willis Hicks PT-17 Stearmans N62917 for a local flight at Chino. This is one of the aircraft that Tony Farhet built at Vintage Aeroplane Company.

In 1980, Hartlay established the flying demostration team, "Silver Wings," made up of three PT-17 Stearmans one of which has a wing walker. The team consists of Hartley flying lead with Margret Stivers as the wing walker and Jimmie New and Ron Caraway flying the other two aircraft. In the ten years that the "Silver Wings," team have been together, they have performed at more than 75 airshows around the country.

In addition to the airshow work, Hartley's Stearmans have been used in television commericals and most recently in a music video for the movie "Mission Impossible II," performed by the rock group "Metalica."

Before and after photos of The Air Museum - Planes of Fame, Stearman PT-17 / N61445, which was built for Walt Disney's movie "The Kid."

One of the nicest looking PT-17 Stearmans operated at Chino is Mac McCauley's Red N450WT. This aircraft was involved in a minor taxi accident and it was amazing to see a group of other Stearman owners come together and help dismantle and repair the aircraft to get it back in the air as soon as possible. (Photo by Philip Wallick)

The other Willis Hicks PT-17 Stearman N5085N, shown taxiing carries the U.S. Army Air Corps training colors, like the Stearmans that were operated at Cal-Aero Field during the war years.

A flight of Stearman N2S-4's near Chino Airport. The aircraft in the break was being piloted by Steve Gibbs, who purchased the Stearman from Dale Clark. The lead aircraft was Bill Klaers of WestPac Restorations. Based in near by Rialto, WestPac, are well known for being the premier Republic P-47 Thunderbolt restoration company. (Photo by Philip Wallick)

George Wildly is another owner at Chino that flies his Stearman as much as he can. The aircraft is painted as a Naval version of the PT-17, an N2S-4 registered as N7122T.

6

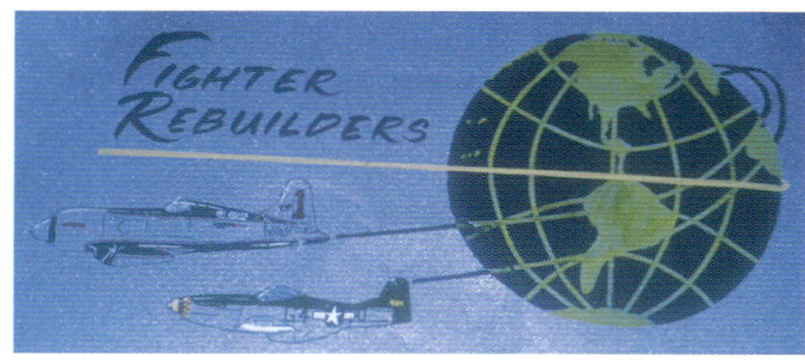

Jim Maloney and Steve Hinton knew each other since the second grade and grew up together around airplanes. They both acquired their pilots license when they were eighteen and soon after started working for Leroy Penhall at Fighter Imports, Inc. in the early 1970's. While gaining experience at Penhall's operation, they both helped restore aircraft for Jim's father, Ed Maloney, at The Air Museum - Planes of Fame.

Then, when Leroy Penhall was killed in the aircraft accident in 1974, Jim and Steve continued to work together on different projects and both flew a number of racing aircraft projects, the "Red Baron RB-51" and later the "Super Corsair." They were also involved in flying aircraft for the television and movie industry, all the while still restoring aircraft at The Air Museum – Planes of Fame.

The shop at "Fighters" usually has one or two aircraft under some sort of reconstruction. Fighters is currently operating out of what was once used as an engine shop in early years when "Pacific Aeromotive Corp," operated at Chino.

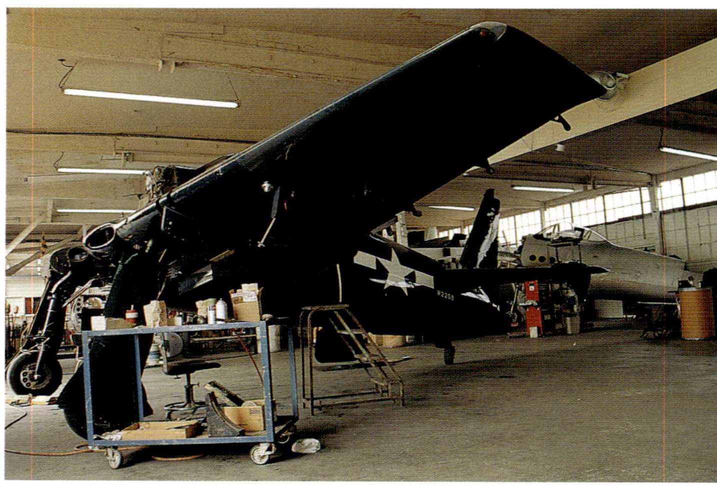

Realizing flying and rebuilding aircraft would always be part of their future, Jim and Steve established "Fighter Rebuilders" as a full time operation in 1980. Tragedy struck three years later when Jim Maloney was killed in an aircraft accident in Arizona. It had only been a few years since Steve had survived the near death crash of the "Red

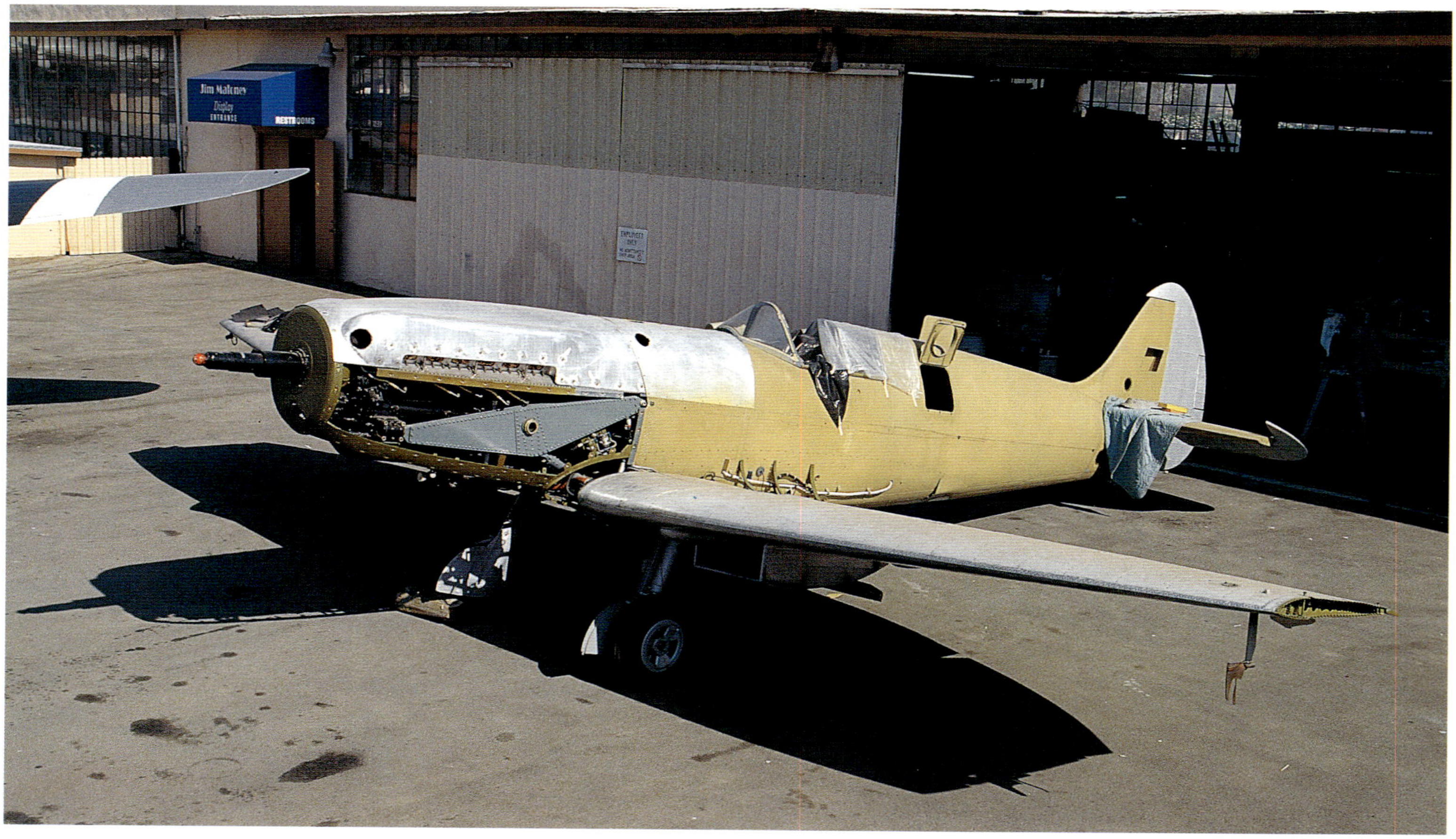

The Air Museum's, Supermarine Spitfire XIX, PS890, shown undergoing a complete restoration at Fighter Rebuilders. Unique to this aircraft are its contra-rotating propellers, that have not been installed in this picture.

Baron" in 1989. Steve was seriously injured again in the crash of a replica Miles Atwood racer while making a test flight at Chino. Amazingly, he would recover from these injuries and just three years later he would fly his stock P-51D at the Denver Air Race in 1996.

"Fighter Rebuilders" has continued restoring aircraft for customers around the world, having established themselves as one of the premier aircraft restoration facilities. "Fighters" as there're known to the locals, has produced well over thirty-five ground-up restoration projects, a number of partial projects and also provides specialized aircraft maintenance for private owners.

A ground-up restoration can take from one to five years to complete. Consideration is given to the condition of the aircraft prior to the start of the project and the expected finished project. The most time consuming task is the fabrication of parts that are unavailable anymore. These have to be hand made to precise specifications to complete the project. Another consideration is the extent of authenticity that the owner wishes the aircraft to be restored to, which could be full military complete with war time paint, or specialized to the owner's personal wants.

As long as there are aircraft projects, "Fighter Rebuilders" intends to continue restoring unique and rare World War II aircraft to their original flyable condition.

At any given time you can find a number of different types of aircraft in different stages of construction at "Fighters." The North American P-51B Mustang was being dismantled for shipment to its owner overseas, while the Grumman C-1 Cod was in for an annual maintenance inspection.

Following is a list of most of the aircraft to have come out of "Fighter Rebuilders":

Aichi D3A2 VAL / NX67629
Bell P-39Q, SN - under restoration
Bell P-63, SN42-68864 / N163BP
Canadair T-33 - RACF 21341 / N12422
Curtiss P-40C, SN41-13390 / N80FR
Curtiss P-40N / N85104
Douglas B-26, SN44-35721 / N9425Z
Douglas AD-5W, BU135188 / NX188BP
Douglas SBD Dauntless / N670AM
Grumman F7F-3P, BU80425 / N7235C
Grumman F7F, BU80412 / N7628C
Grumman F6F-5, BU94473 / N4964W
Lockheed P-38J, SN42-67543 / N3145X
Lockheed P-38J, SN44-23314 / N29Q
Lockheed T-33A / N12413
Mitsubushi Zero - two different aircraft
North American T-6 / SNJ,
    SN42-86287 / N3375G
North American TB-25N,
    SN44-86747 / N8163H
North American P-51A,
    SN43-6251 / N4235Y
North American P-51C,
    SN43-25147 / N51PR
North American P-51D,
    SN44-73053 / N?
    SN44-72051 / N68JR
    SN44-73339 / N51RR
    SN44-74976 / N98582
    SN44-84753 - N251BP
    SN44-84961 - N7715
    SN44-84962 – N9857P
    SN45-11367 - N4078K
    SN45-11582 - N5441V
North American T-28B / N28BP
North American F-86F
    SN52-5139 / N86F
Republic P-47D-4, SN45-49025 / N47RP
    SN45-49192 / N47DD
Republic P-47G SN42-25234 / N3395G
Supermarine Spitfire XIX, PS890 - N?
Vought FG-1D, BU92529 / N62290
Vought F4U Corsair / N83782
Vought F2G Corsair / N3151B
Yak-11

The downfall of the Soviet Union has brought a large number of Warbirds to the open market for collectors. One of the many projects that came out of Fighter Rebuilders was this Curtiss P-40C, SN41-13390 / N80FR, flying after a complete rebuild of an aircraft that came out of the former Soviet Union, where the aircraft had crashed during WWII. Though these aircraft need extensive rework, at least there will be more World War II aircraft available for restoration projects around the world. The P-40 is being chased by The Air Museum's Mitsubishi A6M5 Zeke 61-120 / N46770. The 99% authentic Zero is most impressive and the only flyable original in the world.

Fighter Rebuilders operates from one of the oldest buildings on the famous World War II CAL-AERO Field. In its early years, the building served as an engine overhaul facility and was used for aircraft parts storage. The building is now the primary location where aircraft restoration projects take place.

Fighters also works for The Air Museum as a contractor, supplying a full-time maintenance crew for all the flyable aircraft operated by The Air Museum and restoration of new Warbirds for The Air Musuems's. The restoration area also provides a working display for visitors of The Air Museum to watch the reconstruction process. Fighters is also involved in operational maintenance programs that help keep many vintage aircraft flyable for private collectors and museums around the world.

Two of the more interesting projects taken on at Fighters, were highly competitive air race airplanes, designed only to be raced. The first aircraft that came from the shop at Fighters was the highly modified Vought F4U Corsair, known as the Super Corsair, powered by a Pratt & Whitney R4360. This aircraft was raced for over ten years.

Another aircraft that was a complete construction project, was "Tsunami," for Jack Sandberg, strictly as an unlimited race plane. The aircraft was powered by a Merlin engine.

Fighters, also made improvements and modifications to the "Pond Racer," a revolutionary twin engine race plane designed and built from composite materials.

"Figthers" built Lockheed P-38J, SN44-23314 / N38BP for Bob Pond as "Joltin Jose". The aircraft sat in storage for years at The Air Museum at Claremont, the Ontario facility and later at Chino before being restored. (Photo by Gerald Laing)

Probably the nicest P-51 that was built at "Fighters" was for Jack Sandberg's North American P-51D, SN44-72051 / N68JR "Platinum Plus." (Photo by Jerry Wilkins)

A total of three Mitsubishi Zero replica airframes where scratch built by an Eastern Block country and shipped to the United States by a private owner. "Fighter Rebuilders" completed two of these projects into flyable aircraft. Some modifiactions had to be done to the airframe to accomodate the Pratt Whitney 1830 radial engine, fabrication of a new engine cowling, complete cockpit instrumentation, avionics package and complete rewiring of all the electronical system.

"Fighters" did a ground-up restoration on this Bell P-63 "Kingcobra" SN42-68864 / N163BP for aircraft collector Bob Pond. The aircraft is based at the Palm Springs Air Museum, Palm Springs, California. Ground up reconstruction projects can take years to complete and as you would expect, a lot of money!

7

AERO TRADER

Warbird Restoration & Maintenance

Aero Trader was established in 1976 when Carl Scholl and Joe Davis became partners, after purchasing Jack Hardwick's large supply of Warbird parts. Then in 1979, Tony Ritzman, bought into Aero Trader as a third partner and later still in 1987, Carl and Tony bought out Joe's share of Aero Trader. Carl Scholl and Tony Ritzman became partners on a North American B-25 as a way of going to airshows and enjoying their interest in aviation. Carl had already purchased an aircraft years before and had begun restoring aircraft as a professional hobbyist. Over the next few years, Carl and Tony flew for the movie industry.

Aero Trader opened its shop at the Chino airport in 1985, when they realized that more room and better working areas were needed to produce a quality product. Aero Trader has the reputation of building the highest quality, most authentic North American B-25 Mitchells in the world!

Photos on both pages: The beginning of and almost completed project. At any one time, there are at least two North American B-25 Mitchells being completely rebuilt at Aero Trader, while others undergo some sort of maintenance. Aero Trader has won a number of awards for authenticity, craftsmanship and quality on the B-25's that have emerged from their hangars.

Currently, Aero Trader is the largest privately owned Warbird facility in the United States, if not in the world. They operate with a full time staff of over twenty people, operating out of two large hangars with more than 20,000 square feet. Each is large enough to assemble three complete B-25's. The facility has a complete machine shop and other necessary facilities to fabricate anything that might be needed to complete a restoration project. The Chino facility has a number of storage containers located in the back of the facility that house parts that are being used, or parts that may be needed to complete projects underway.

Aero Trader operates a plexiglass shop located on the Chino Airport that produces all glass associated with the B-25, including the domed gun turrets. They also make P-51 front windshields, side panels and other special order items.

Aero Trader also operates a parts warehouse for the B-25. They currently have more B-25 parts than any other supplier in the world. Not many people are allowed out to what Carl calls his secret desert storage yards, where thousands of parts are stored. There are quite a few complete aircraft awaiting restoration at the Southern Califorian desert facility which also has a runway.

Only six months into World War II, the United States had to prove that they could strike anywhere at any time. General Jimmy Doolittle lead a flight of B-25's from the aircraft carrier U.S.S. Hornet on a bombing raid in the heart of Japan.

To commemorate the 50th anniversary of the raid, three North American B-25's were flown off decks of the U.S.S. Ranger - CVA-61, while in southern California waters near San Diego. Carl and Tony participated in this reenactment by flying "Pacific Princess," off the deck of the U.S.S. Ranger. Prior to the event, the B-25 aircrews had to prove to the U.S. Navy that they could takeoff in a designated distance that was marked off on a hard surface runway at Naval Air Station North Island, San Diego, California. After the completion of takeoffs on dry land, the three B-25 air crews were approved to fly off the flight deck of the aircraft carrier U.S.S. Ranger while at sea.

Then, later in 1995, the "Pacific Princess" was loaded aboard the U.S.S. Carl Vinson, CVN-70, where it was flown off the flight deck to participate in the U.S. Navy's annual "Fleet Week" activities and airshow held at the Naval Air Station at Barbers Point, Hawaii.

Photo above, "Pacific Princess" was the first North American B-25J, SN44-30832 / N3155G rebuilt and owned by Carl Scholl. The aircraft is still flown at airshows and in motion pictures.

The hangars at Aero Trader have seen a number of different types of aircraft. A variety of Douglas B-26 Invaders have been rebuilt, while others underwent annual inspections, normal maintenance, or were partial restoration projects.

This page contains some of the nose art and colorful markings adorned on some of the aircraft that have passed through the Aero Trader hangars. Some of the aircraft projects leave unpainted while others have complete military style paint schemes applied as part of the project.

Not a total secret, but Aero Trader has a large storage facility at Ocotillo Wells, California. The facility is complete with a runway long enough to operate a B-25 from. The amount of aircraft, parts, engines and other aviation related equipment continually changes as aircraft are bought and sold. Some of these projects are taken to the Chino facility and rebuilt, while others remain in the California sun. Following is a list of the North American B-25 Airframes in storage at the Ocotillo Wells, California desert storage faciltiy: SN40-2347, SN42-32354, SN43-27596, SN44-28738 / N3441G, SN44-28765 / N9443Z SN44-30090 / N9633C, SN45-8887 / N3680G SN44-30761 / N3398G

Looking at these pictures you can see that there are some very unique aircraft; parts of two Martin B-26's, a fuselage of a Douglas A-20, a complete Boeing B-29 and parts of a second in storage, CA-26 ex-Royal Australian Air Force Sabres, aircraft round and inline engines, landing gear and more even a BT-13. There are also some aircraft stored in large shipping containers for private owners.

Aero Trader is the best at restoring B-25's, but they have also rebuilt a variety of other types of aircraft with the same quality. The following is a listing of the aircraft that Aero Trader has completed over the years:

Commonwealth CA-27 Sabre
Curtis P40N, SN44-7983 / N9950
Douglas B-26C, SN44-35710 / N770SC
Gnat T. 1 XR955
Martin B-26, SN40-1501 / N4299S
Mig-15'S (2) / N7103L & N7103N
North American T-28C, BU140511 / N140AG
North American P-51D in restoration

North American TF-51 A68-198 / N286JB
   Conversion aircraft went to France
North American;
   B-25D, SN44-3318 / N88972
   B-25J, SN44-30832 / N3155G
   B-25J, SN43-28204 / N9856C
   B-25J, SN44-86797 / N3438G
   B-25J, SN43-28059 / N1943J
   B-25J, SN44-28925 / N7687C
   B-25J, SN45-8835 / N5672V
   B-25J, SN43-28059A / N1943J
Vought F4U-5NL Corsair
   BU124447 / N100CV
   FG-1G Corsair, BU67089 / N96GM

Though the B-25 is the main product of Aero Trader, they have also rebuilt and currently maintain a number of other types of aircraft. The CA-18 /P-51D, A68-198 / N286JB was modified with dual flight controls before leaving for its new home in France, as F-AZIE. A number of North American T-28's have been rebuilt or have had annual inspections at Aero trader. The hot rod of the bunch is Dick Berta's Grumman F7F-3 Tigercat, BU80483 / N6178C which was maintained at Aero Trader.

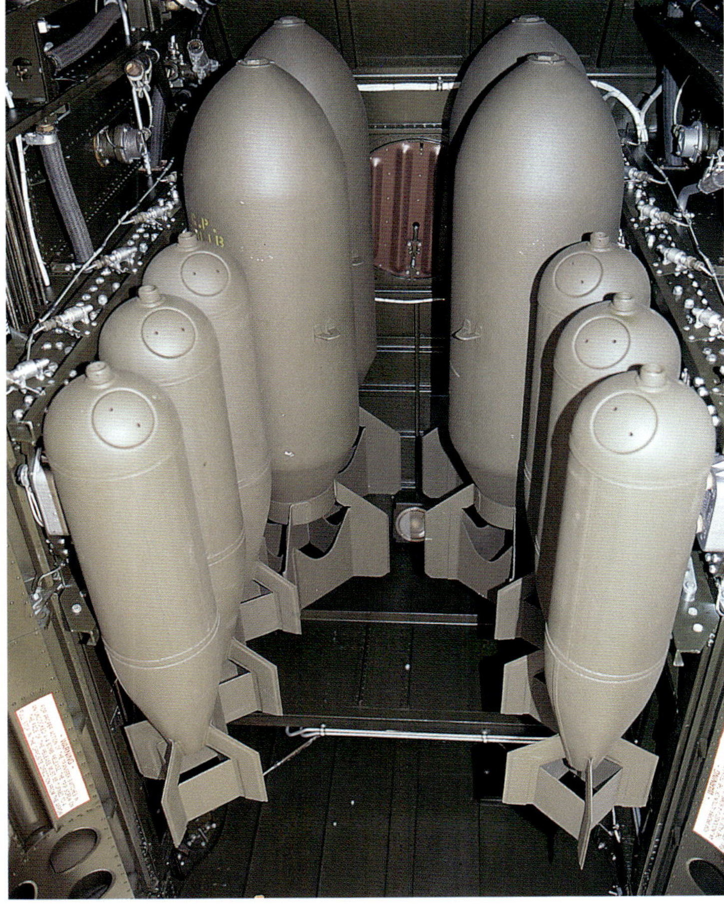

The Cavanaugh Air Museum's B-25 is one of the only flying Mitchells with actual WWII combat time. Every detail of the aircraft was rebuilt as close to the original North American standards, including a Norden Bomb sight.

8

Pioneer Aero was started by Elmer Ward with the intent of providing clients the highest quality standards of restorations of North American P-51 Mustangs. One of Elmer's sons, Bret, ran the company, which still operates as the leader in North American P-51 Mustang parts. In the beginning, the company built a number of Mustangs and grew into their speciality of building TF-51's, dual control Mustangs. They also stock more Mustang parts than any other parts house in the world. The shelves at "Pioneer"are full of the most common to the most obscure parts. Pioneer had established themselves as the premiere company building the most sought after warbird in the world, the North American P-51 Mustang. Pioneer had built four Mustangs and the Gulfhawk Bearcat. Three of the Mustangs were built for Warbird collector Doug Arnold of England. These aircraft were; "Petie 2nd" P-51D SN44-73140 / N314BG, "Shangrila" P-51D SN44-72934 / N513PA and the first TF-51 SN44-73871 / N7098V built by Pioneer. Doug Arnold's collection has since been scattered around the globe after his death. P-51D SN44-73435 / N6311T was the fourth Mustang built at Pioneer Aero. P-51D SN44-63507 / N6345T was the second and last TF-51 built at Pioneer Aero as SN41-406 / N973 "Little One III," this aircraft was later finished at "Square One."

The conversion to a dual control TF was fabricated on engineering that had already been done by Cavalier Aircraft Corporation. Pioneer Aero had the forward fuselage of the TF-Mustang built by Cal Pacific Airmotive, located in Salinas, California. The TF is basically a "D," with a complete set of instruments and flight controls useable from the rear seat. It also has an enlarged canopy for more head room and visibility.

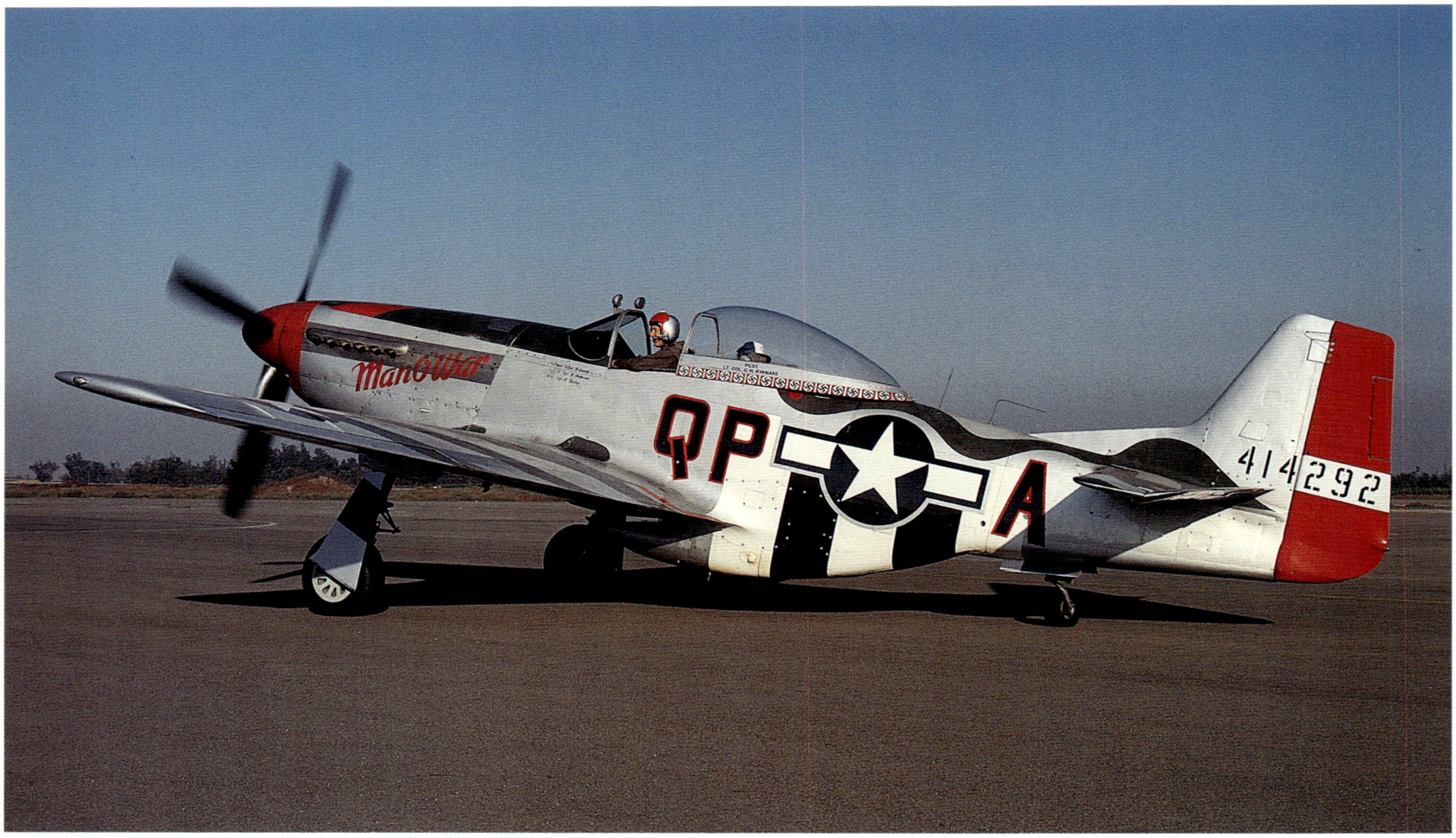

Elmer Ward's North American P-51D Mustang flies as "Man O War" SN41-4292, but is actually SN44-722739 / N44727. This Mustang was originally rebuilt at Aero Sport in the early 1970's and has since been overhauled by Square One, who still maintain the aircraft. Elmer and his son Todd fly the aircraft quite regularly to airshows around the country.

Two of the three Mustangs built by Pioneer Aero, where for aircraft collector Doug Arnold, this were "Shangrila" P-51D, SN44-72934 / N513PA and a TF-51, SN44-73871 / N7098V.

Pioneer Aero was split into two companies in 1995. Pioneer Aero continuing to be the leader in providing North American P-51 Mustang parts, while Square One, the new company, provided maintenance service and retained the full restoration capabilities of all types of Warbird aircraft. The shelves at Pioneer are full of the most common to the most obscure parts. Whatever kind of Mustang part you may need, Pioneer Aero should have it, if not, their sister company, Square One, has the capability of fabricating it.

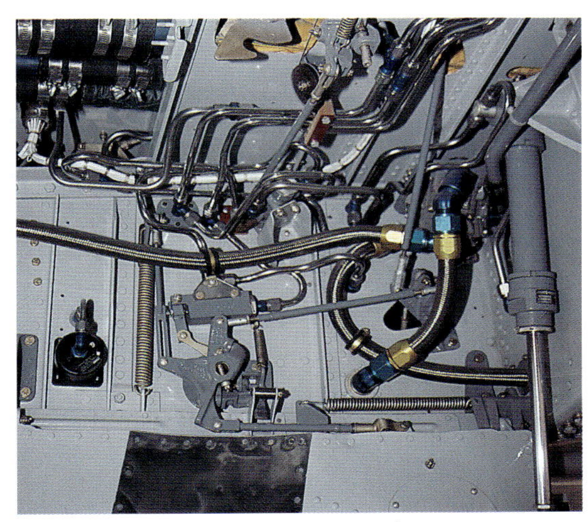

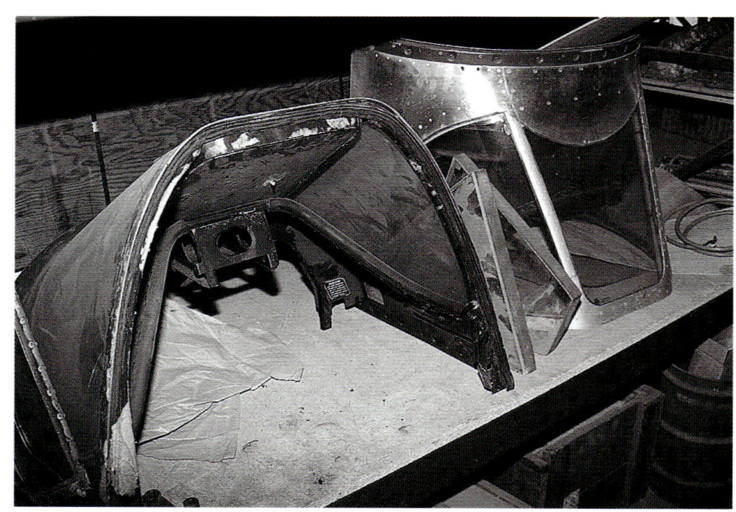

These pages show just a fraction of the items Pioneer Aero has on stock; gun sights, wheel brake assemblies, thousands of small parts, front canopy windshields and new windshield glass, tail wheel assemblies, complete engines and engine parts, canopy frames, oil cooler fairing scoops, elevator and wing flaps, fuel tank cells and drop tanks, landing gear doors, magnetos, main landing gear assemblies, complete fuselage and wing sets and much more that isn't pictured. If you need Mustang parts - Pioneer Aero is where you'll find them.

Grumman G-58A "Gulfhawk 4th"
F8F-2 BU121707 / NL3025

Major Alford Joseph Williams Jr. became famous during the 1920's through the 1940's for his airshow demonstrations. The first aircraft Williams used was a Curtis Hawk, the original "Gulfhawk," then the most famous, a Grumman F3F – called "Gulfhawk 2," which is currently at the Smithsonian National Air & Space Museum. "Gulfhawk 3" was one of only 2 civilian versions of the Grumman F8F Bearcat, designated the G-58A. All aircraft were painted bright orange with white and blue stripes. The Gulfhawk 3, G-58A didn't last long, only four months. During an emergency landing, the undercarriage failed and the auxiliary fuel tank split open causing a fire which engulfed and destroyed the aircraft.

Elmer Ward had owned a P-51 Mustang for a number of years and wanted a Bearcat. A project aircraft was purchased in 1982 and 11 years later "Gulfhawk 4" emerged from the hangars at Pioneer Aero. The wing box and spar were strengthened and wings were completely built from scratch. All copper and aluminum tubing was replaced with stainless steel. All spot welds were replaced with rivets for added strength. The aircraft was modified to carry two people and basic hardware for dual flight controls was installed, but not completed. The back seat was raised a little for better visibility and the canopy was now 12 feet long. Another major change was to make the cockpit instrument panel the same as the P-51 Mustangs that Pioneer Aero had built.

When finished, the aircraft was a beautiful accomplishment. After two test flights at Chino, the aircraft was flown to the 1992 Oshkosh Airshow. A year would pass and about 80 flight hours were put on the aircraft taking it to Airshows in California before going back to Oshkosh in 1993. This would be the last place the aircraft would visit. Upon departure from Oshkosh, the aircraft crashed due to lack of power during take-off. Elmer was very lucky and survived with only minor injuries.

In theory, the aircraft was a complete write-off, but because of the technology of today's Warbird manufacturing knowledge and expertise, the aircraft could be rebuilt and might fly again!

GULFHAWK Jr.
NL3025

9

The summer of 1995 saw another change in the Warbird manufacturing industry at Chino. Elmer Ward's well established Pioneer Aero split into two companies. Pioneer Aero would remain the premiere P-51 parts house while the new company, Square One, would become the manufacturing and maintenance facility. The new facility is located next to Aero Trader on the south row of large hangars on the western part of the airport.

Elmer Ward's business philosophy of Pioneer Aero would carry over to Square One, that is "Quality at its best!" The company became famous for building some of the highest quality Mustangs flying today. The conversion jig to fabricate the "TF" fuselage came from original "Temco" design drawings and has become a very popular modification. All fabrication is done by highly qualified and experienced Warbird mechanics.

At any one time there are two or three Mustangs in the hangar undergoing either full restoration projects, or just in for scheduled maintenance. The aircraft that emerge, from the hangar at Square One are as good, if not better than those that rolled off the North American Mustang production lines in war years. This holds true whether they are Mustangs or another type Warbird.

Elmer's, Mustang, "Man O War," would find its new home at Square One new hangar, where it continues to receive a military style preventative maintenance program. Mustangs still seem to be the dominant aircraft being produced at Square One.

The first project out of the hangar at the new Square One facility would be, of course, a TF-51 – N6345T, built as "Little One" which flew with the 352nd Fighter Group in Europe. Sad to say, this aircraft crashed and was a complete write-off only months after its completion. Another TF-51 was built from the remains of ex-racer #6 "Somthin Else" / N51VC and remains of another Mustang. This aircraft was built for B.L Jennings / N1451D – "Saturday Night Special." The aircraft carries a non-military looking, yet impressive, extremely highly pol-

The shops at Square One usually have two or three Mustangs in different stages of construction. The TF dual control Mustang is their speciality, but they also build other types of warbirds.

One of the many stages of a restoration project.

ished aluminum finish. Ex-Astronaut, Col. Frank Borman had Square One build a TF-51 from pieces of two Mustang airframes to make – N50FS – "Su Su II." After being retired from air racing, the highly modified "#84 Stiletto" had its fuselage converted to a TF-51. The complete aircraft – N332 was rebuilt back to a military style configuration with the aircraft rebuild finished at another facility as the "Rajun Cajun." Other Mustang projects include fuselage conversions of stock single seat "D" model aircraft into dual control two seat TF-51 Mustangs. One of these is the second TF-51 Mustang flown at "Stallion 51" Corporation, named "Mad Max." Two other "TF" fuselage conversions have been built for other clients around the country. The company has also started work on a TF-51 – N151SQ that will be used as a demonstrator aircraft to allow prospective clients the opportunity to fly a finished product before committing to the dollar investment.

Square One has departed from just solely rebuilding and manufacturing Mustangs. This will be done with the same high quality craftsmanship they are known for producing -51's, only now it will be on any type warbird, including the Jet aircraft community. The first non-Mustang product was the beautiful Bell P-63 Kingcobra – N163FS rebuilt for Col Borman. The aircraft was a rare find, having been stored since 1946 with only a limited number of flying hours on the aircraft. The first jet projects would be two L-39's and an F-86.

A final word about the remains of Elmer's, F8F-2 / G-58A "Gulfhawk," which was completely destroyed at Oshkosh during a take-off accident. The aircraft was returned to Chino soon after the accident. Fighter Rebuilders was contracted to rebuild the center section of the Bearcat, while Square One started rebuilding other parts of the aircraft. The remains of the aircraft are in storage at Square One and it is not known what the final outcome of the aircraft will be.

Frank Borman's P-63 Kingcobra shown stripped down during reconstruction at Square One. When the aircraft was located, it was found in a T-hangar on the Van Nays airport. In 1946 the aircraft made an emergency landing which caused all three tires to blow and the aircraft was abandoned until Frank Borman bought the aircraft in 1996. Considering that it had sat for almost fifty years, it was in very good condition. After being rebuilt it won WWII Grand Champion at the 1998 Oshkosh Airshow.

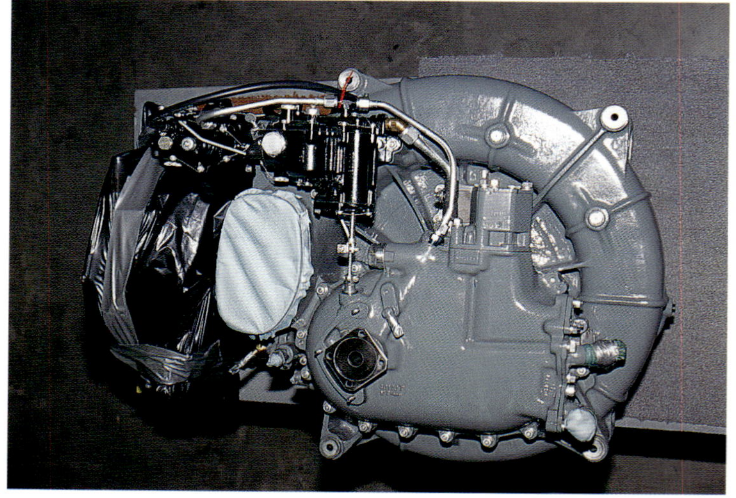

The P-63 is powered by an Allison 12 cylinder V-1710 turbo supercharger engine which was overhauled by "Nixon's Vintage V-12," in California. The Allison produced 1,800 horsepower and powered the aircraft to speeds around 400 miles per hour.

P-51D SN45 11525 / N151AF "Val-Halla" is owned an operated by ex-Astronaut Bill Anders, who also owns a Grumman F8F Bearcat and two North American T-28's. The P-51 was at Square One having an annual inspection completed.

It's not uncommon to see one or two over-hauled Merlin power plants on stands in the Square One hangar. The Packard-Merlin V-1650-7 12-cylinder 1,649 cubic inch inline engine is rated at 1,490 horsepower on take-off.

The following is a list of aircraft that have been produced or are under construction at Square One:

P-63   - SN42-69021 / N90805
F-86E - SN55-71461/ N186FS
P-51D - SN44-63655 / N5500S
P-51D - SN44-63865 / N51JK
TF-51 - SN44-63507 / N6345T
TF-51 - SN44-74446 / N1451D
TF-51 - SN45-11471 / N332
TF-51 - SN44-73832 / N151SQ
TF-51 - SN44-73117 / N251SQ
TF-51 - CAG-18 / A68-187+
        SN44-74839 / N50FS

Another one of Frank Borman's aircraft is this North American F-86E SN55-71461/ N186FS. This aircraft was once owned by Micheal Dorn of "Star Trek" fame.

10

Charles Nichols originally started collecting United States military aircraft in 1975 in Compton, California as a hobby. Then, by 1982, the collection became too large to be just a hobby - it became known as the "Yankee Air Corps Museum" and was moved to the Chino airport. Four large hangars were constructed, two of which housed the museum's aircraft, while the other two where leased out. The collection of aircraft continued to grow and in 1994 another move was made, this time to an even larger hangar where all the aircraft could be displayed in one area. This new facility was built adjacent to the museum's maintenance hangar where all restorations are currently done. The move also promoted a new name, "Yanks Air Museum."

The museum is currently located on the south-western part of the Chino airport and continues to restore aircraft to their original flying condition. The restoration hangar is one of the largest hangars on Chino Airport capable of housing quite a few complete aircraft. During the Vietnam Conflict, the hangar was used to manufacture napalm (jellied gasoline) for the military. The staff at "Yanks Air Museum" is large enough that they can work on a number of different projects at one time. There is also a large storage yard located in back of the maintenance facility were a number of aircraft are awaiting their turn to be rebuilt. The museum specializes in U.S. Military fighter, observation, trainer, utility and a few civilian rare or unique aircraft. Most recent additions are the jet era collection.

There are also a number of vehicles and artifacts displayed around the aircraft and on the walls at Yankee. Like all museums they have a continuing search for more aircraft and war artifacts that could be added to the collection and displayed as historical information for the public.

Photos starting opposite page top - part of the jet aircraft collection are the North American FJ-1 Fury and Lockheed P-80 Shooting Star both aircraft are waiting restoration. Located inside are a Pre-World War II Trainer, the Stearman YPT9B, a Naval Aircraft Factory N3N-3 is a construction cut away display and the museum also houses a very rare N3N-3 on floats. Part of the World War II collection consists of a Curtiss P-40 Kittyhawk, North American P-51D Mustang and Douglas SBD Dauntless.

Yankee has a number of aircraft that are rare and unusual, one of them is this Lockheed F-5G Photo Lighting. At one time this aircraft had been modified with a pressurized cockpit.

The Grumman TBM Avenger is complete and waiting its turn for restoration. Sitting next to the TBM is a very rare Vought OS2U Kingfisher that is undergoing a complete restoration.

Yankee's Bell P-39 Airacobra is another aircraft that was acquired from the Military Aircraft Restoration Corporation and has been completely rebuilt.

The Curtiss SB2C Helldiver was acquired from The Air Museum a number of years ago and has been in storage since. It will be a nice addition to the collection of naval aircraft at Yankee when completely restored.

The North American B-25 Mitchell was recently acquired from Australia. The aircraft was part of the Australian War Memorial until Yankee purchased the aircraft.

Also displayed at Yankee, is a Curtiss O-52 Owl, used by the Army Air Corps for observation, liaison, trainer and utility duties. This aircraft was acquired from the Military Aircraft Restoration Corporation, they also have another complete O-52 airframe in storage that could be restored.

The North American P-51A is one of only a few of its type displayed in museums around the country. They also have a "D" model Mustang on display.

Yankee, has two Republic P-47 Thunderbolts, a "D" model and a very rare "YP-47M."

The Bell P-63 Kingcobra is also one of only a few of its type either displayed or still flying. Yankee also has a Bell P-39 Airacobra in the final stages of restoration.

YANKEE AIR MUSEUM
AIRCRAFT LISTING

Beech UC-18 / D17S Staggerwing
SN43-10842 / N51746
Bell P-39N Airacobra SN42-8740 / N81575
Bell P-63C Kingcobra SN42-69080 / N94501
"Fatal Fang"
Bell 47D1 Helicopter SN114175
Consolidated BT-13 Vultee SN79326 / N4425V
Curtiss SB-2C Helldiver BU19075 / N4250Y
Curtiss P-40E Kittyhawk AK827 / N40245
Curtiss O-52 Owl SN40-2769 / N61241
Douglas SBD-4 Dauntless BU10518 / N4864J
Grumman FJ-1 Fury BU120349
Grumman TBF-1-C Avenger BU05997
Grumman F8F Bearcat BU122095 / N2209
Grumman F6F-5 Hellcat BU78645 / N9265A

Grumman F-14A Tomcat (2 - airframes)
Kellett YG1B Autogyro SN37-301
Lockheed P-80A Shooting-star
Lockheed F-5G Photo-Lighting SN44-27183 /
N517PA
Lockheed 10 Electra
McDonnell-Douglas F-4D Phantom
Naval Air Factory N3N-3 N45265, N45280
80% complete aircraft skelton cut
away airframe on display.
North American P-51A Mustang
SN43-6274 / N90358
North American P-51D Mustang
SN44-74910 / N51SJ
North American SNJ-5 Texan
SN41-31532 / N43771
North American B-25 Mitchell
Republic P-47D Thunderbolt
SN45-49346 / N3125D
Republic YP-47M Thunderbolt
SN42-27385 / N4477M
Republic F-84 Thunderstreak
Sikorsky H-3 Jolly Green Gaint
Stearman YPT9-B / N795H
Stearman 4D / NC11224
Vought F4U Corsair BU97390 / N47991
Vought OS2U Kingfisher
Western FM-2 Wildcat BU86564 / N4629V

11

During World War II, military aviation was a decisive force in U.S. military actions around the world. More than 300,000 warplanes were produced in America during the war years, but soon after the war they were almost scrapped to total extinction. After the war only a small percentage of the the massive U.S. Military air armada survived the cutting torch. Cal-Aero Field, which is now Chino Airport, contributed to the demise of a lot of those aircraft. Just after the war, thousands of military aircraft were being sold or scrapped. More than 5,000 aircraft were scrapped at the government depot on Cal-Aero Field. The government did have a program to preserve only a few aircraft for historical purposes. These aircraft would be used for display at future aviation museums around the country, like the Smithsonian Air and Space Museum, the U.S. Air Force, U.S. Army, U.S. Marine Corps and U.S. Naval Air Museums.

Ed Maloney established The Air Museum in Claremont, California on January 12, 1957 as the first permanent aviation museum west of the Rocky Mountains. Interested in all aspects of aviation, especially the "Warbirds," Ed was shocked to see so many aircraft used during World War II being reduced to scrap metal after they returned from the battle fields of the war. At this point the concept was simple, save as many of these aviation historical treasures as possible.

"The Air Museums" maintains more than thirty aircraft in flyable condition, these are some of those aircraft: N9MB Flying Wing, F-12E/F4B3, P-26 Peashooter, B-25 Mitchell, F-86 Sabre, Mig-15, Ha 1112/Bf-109, A6M5 Zero-Sen, P-51 Mustang, F4U Corsair and F3F. Some of these aircraft are the only one of its type flyable in the world!

This is what The Air Museum looked like when it was located at the Ontario Airport during 1963 thru 1970. Most of these aircraft would have been scrapped by the government if not for Ed Malloney. (Both photos this page by Frank Mormillo)

The Air Museum during the early years at Chino hosted airshows on a regular basis. The main structure in the center is the old engine shop and is currently being used by Fighter Rebuilders.

The Air Museum utilizes ex-CAL-AERO hangar #4, as their Jet and Air Race Museum. Some of the aircraft on display are the "Schneider Trophy Air Race" seaplanes and sport air racing type aircraft.

This very rare Culver PQ-14 Target Drone is one of the aircraft displays. There are many types of aviation artifacts and photographs that are all part of displays at The Air Museum.

The Fighter Jets Museum has a varitety of aircraft on display, from the Korean War to Vietnam, these include, a Soviet Mig-15 in North Korean markings, North American F-86 Sabre and Vought F8U Crusader.

A variety of Pre-World War II U.S. aircraft are on display in the North Hangar. There are also many displays of plastic models, aircraft engines, military artifacts and aviation photographs.

The Air Museum also has a large collection of different World War II German aircraft displayed in the North Hangar.

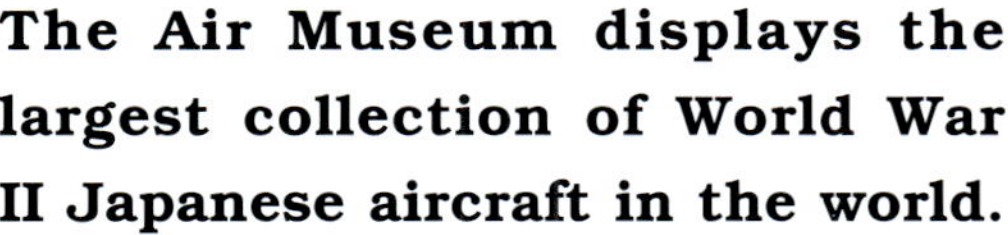

The Air Museum displays the largest collection of World War II Japanese aircraft in the world.

In the early years, Ed Maloney collected as many aircraft and aircraft parts as he could, storing them in his backyard. Supported by his wife, Louise, and their children, Ed Maloney took a major step in his quest in the preservation of aviation history. In January 1957, he opened a small aviation museum in Claremont, California, known as The Air Museum. Continually looking for aircraft at government surplus auctions, public aircraft auctions, trade schools and the few private owners that had acquired aircraft. The Air Museum has since grown into one of the largest privately owned and operated aviation museums in the United States, if not the World.

Some of the aircraft Mr. Maloney obtained were only parts of aircraft, or had been damaged to the point where they could only be used as static displays. Others were pieces of a number of different aircraft used to make one complete aircraft that could be displayed. There were also a lot of different aircraft parts that were obtained with the hope they could be used on future aircraft projects, as they became available. As years went by, some of the parts were utilized, while others were traded off for things that the museum could use. Even today, The Air Museum is continually looking for more aircraft and artifacts to improve and expand the collection of aviation history they display.

The Air Museum has moved a number of

times during its existence and in the process many improvements occurred, but today "The Air Museum - Planes of Fame" is located at the famous WWII Cal-Aero Field, now known as Chino Airport. The Air Museum - Planes of Fame, not only preserves aircraft for display, but houses some of aviation's most historical aircraft in flyable condition, some of which are the sole surviving examples of their type.

Whether they were record setting air racers or famous fighter aircraft of the past wars, the aircraft on display at The Air Museum were built in a time gone by, but their unforgettable sound and smell brings back memories for those lucky enough to have experienced them during their life. The museum also allows those interested in aviation the opportunity to enjoy a variety of different aviation eras first hand.

The Air Museum, is truly a Warbird sanc

The museum has many different types of aircraft parts and other historical aircraft memorabilia displays. Wooden models and hundreds of plastic models cover every era of aviation history. These were donated by museum members and aviation enthusiasts who have volunteer their time and efforts to preserve aviation history.

There are only two original examples left of the 136 Boeing P-26's that were built. The Air Museum's Boeing P-26 Peashooter is the only flyable example of its type in the world.

tuary, with a number of the aircraft displayed being the only example of its type surviving in the world. Included in this category are the following aircraft: Northrop N9MB Flying Wing, Japanese Mitsubishi A6M5 Zero-Sen Fighter, Mitsubishi J8M-1 Shusui Rocket-powered Interceptor, Mitsubishi J2M3 Raiden Interceptor, Aichi D3A2 Val Dive Bomber, Boeing P-26 Peashooter Fighter, Boeing P-12E/F4B-3 Pursuit Fighter, Seversky 2PA/AT-12, Ryan FR-1 Fireball Prop-Jet and a German Horton Ho IV Flying Wing Glider.

The Grumman F3F-2 that is diplayed at The Air Museum is one of four aircraft that were manufactured by The Texas Aircraft Factory. These aircraft have a few original parts from many Grumman F3F's that had crashed on the Hawaiian Islands in the late 1930's.

The Air Museum's Boeing P-12, though currently painted as the U.S. Naval version the F4B3, is actually a U.S Army Air Corps P-12E. This is also the only flyable example of its type in the world.

NORTHROP N9M "FLYING WING"

The Northrop N9MB Flying Wing is one of the most significant aircraft that The Air Museum acquired, rebuilt and maintains in flyable condition. It is the only surviving aircraft of the four N9M's that Northrop built in 1942. The four aircraft were proof of concept scale models of the larger "Flying Wing," the B-35 Bomber that was purposed to be built for the U.S. Air Force.

The N9 series aircraft, were built as third scale models of the larger flying wing to prove the all wing-flight capabilities. Instead of conventional tail surfaces, the flying wing uses elevons to control aircraft pitch and roll. Rudder pedals control split trailing edge flaps independently for lateral control. Flight testing, proved that the flying wing had excellent flight characteristics. The four N9M test aircraft flew hundreds of successful test flight hours in the three-year flight test program.

The N9MB was saved from being scraped and sat at The Air Museum for many years before a 13 year restoration project was started. (Photo by: Philip Wallick)

The N9M is constructed with a combination wood and, fabric and metal frame. The wing's outer panels were constructed of wood and covered with veneer paneling. The first two N9's were painted yellow to make it easier to see them from the ground during flight testing. The third was painted blue on top and yellow on the bottom. The fourth, which is The Air Museum's aircraft, is painted yellow on top and blue on the bottom.

The airframe of the N9MB was saved from the cutter's torch and sat for many years before a complete restoration project of the Flying Wing would take place. It took more than 13 years, an uncountable amount of man-hours and a lot of money to get the aircraft back in the air. The "Flying Wing" is maintained and kept flyable by a number of volunteers and flown on a regular basis at airshows and special events.

The Air Museum's Lockheed C-121A "Constellation," nicknamed "Bataan," was General Douglas MacArthur's personal aircraft. (Photo by: Frank Mormillo)

The Air Museum's, Lockheed C-121A is one of only a few still flyable in the World. The aircraft designated was as a C-69 / C-121/ VC-121A / EC-121N "Constellation," serial number 48-8613, had a very unique history having flown as General Douglas MacArthur's personal aircraft, nicknamed "Bataan." It was later converted back to a normal cargo transport of the U.S. Air Force and was the first "Connie" equipped with radar. The aircraft was finally retired from active military service and transferred to the National Air and Space Administration (N.A.S.A) for use in the Astronaut training program during the Apollo Space Moon Missions. The aircraft was again retired and put on static display at the U.S. Army Aviation Museum at Ft. Rucker, Alabama. It was acquired by "The Air Museum," in 1992, made flyable and restored back into the VIP configuration of "Bataan" to include its complete interior, just as it was when flown for General MacArthur. The aircraft was flown back to Chino and is currently displayed at the "Planes of Fame" – Grand Canyon Valle-Williams Airport location in Arizona.

The cockpit of the "Connie" was basic, but state-of-the-art for her day.

The Air Museum's Ford 5-AT Tri-Motor was used for sight seeing at the Grand Canyon for many years and is normally kept at the Planes of Fame - Grand Canyon location.

The Air Museum's Boeing B-17G Flying Fortress, 44-83684 / N3713G carries markings of SN44-83684, "Picadilly Lily," from the television series "Twelve O'Clock High." The aircraft sits as the Museum's front gate guard and though not flyable, it could be restored to fly again.

The Air Museum's B-25J, SN44-30423 / N3675G "Photo Fanny" has been the workhorse for the museum for the last twenty years. The aircraft has been used as a camera platform for movies, television commericals, air-to-air photography and has been flown to airshows around the country.

Boeing B-17G, SN44-83684 / N3713G being flown just north of Chino Airport. This aircraft was used in the television series "Twelve O'Clock High." (Photo by Tom Piedmonte)

The Air Museum's Douglas RB-26C Invader, SN44-35323 is the photo-reconnaissance version of the famous World War II / Korean Conflict medium bomber. This particular aircraft saw combat in the Korean Conflict and is still flown on a regular basis.

The Air Museum's Douglas SBD Dauntless, BU28536 / N670AM has actual combat time while flying with the Royal New Zealand Air Force during World War II. This aircraft is one of only a few flyable Douglas SBD Danutless's in the world. Recently, there have been a number of Dauntless dive-bombers recovered from Lake Michigan, but most have only been made into static displays for museums around the country.   The aircraft is normally painted in standard naval blues, but was painted to make it look like a U.S. Army Air Corps A-24 for the U.S. Air Force's 50th Anniversary.

The Air Museum's Vought F4U-1A, BU17995 / N83782 Corsair while the museum was located at the Ontario Airport. The yellow and red wing in the background is the Museum's Grumman F6F when it was used as a drone by the U.S. Navy. (Photo by: Carl Porter)

There were 12,275 Grumman F6F Hellcats produced during the war years, of those only 26 are known to exist and less than half of those are flyable. The Air Museum's Grumman F6F-5 Hellcat, BU93879 / N4994V has been maintained in flyable condition since it rolled off the production line over fifty years ago.

The Grumman TBM-3, BU91264 / N7835C Avenger torpedo bomber became famous when it was were used to sink a large number of Japanese warships during the World War II.   After its military career, the Avenger was used as an agricultural crop duster and as Fire Fighting Tankers.   There are quite a few Avengers still flying and they can be seen at Airshows around the World.

The Air Museum maintains a very rare Republic P-47G, Razorback Thunderbolt, SN42-25234 / N3395G, shown sitting on the ramp at Chino.

There were almost 10,000 Lockheed P-38 Lighting's built during the war years. The Air Museum's P-38, SN44-23314 / N38BP is shared with "The Palm Springs Air Museum" and is one of only a handful still flyable.

The most famous World War II fighter and a very popular warbird, the North American P-51 Mustang flying in formation with one of the most rugged fighters of its day, the Republic P-47 Thunderbolt. These are just part of The Air Museum's fleet of aircraft that are maintained in flyable condition and flown on a regular basis at airshows and special events at the museum.

The Air Museum's collection of aircraft spans the history of manned flight, from the hang gliders of 1896, to modern jet fighters and even some space vehicle replicas. Some of the aircraft on hand at the museum have come from all parts of the world. The museum continues to expand and add more aircraft as they become available. Today, The Air Museum is recognized as one of the largest and most unique collections of aircraft in the world. Even more important is that The Air Museum has been a major part of the Chino Airport and its historical value to Aviation History, worldwide!

The Air Museum's Curtiss P-40 Warhawk, SN-42-105192 / N85104 carries markings similiar to those flown by General Chennault's "Flying Tigers," while flying in China during World War II.

The first Turbo-Jet powered aircraft to fight in World War II was the German Messerschmitt Me-262 "Schwalbe." This aircraft was one of the many war trophies brought back to the United States for flight test and evaluation, after which it was saved from the cutters torch. This aircraft has since been sold to another aviation collection in North America.

The Air Museum displays and operates a German Messerschmitt Bf-109 as well as a Spanish built Hispano Ha 1112 version of the famous fighter. More than 35,000 Bf-109's were built, of which there are only about 40 known to exist.

The Air Museum currently displays the largest collection of Japanese aircraft in the World. The Mitsubishi J8M1 Shusui, was the first aircraft saved from the scrap yard many years ago by Ed Maloney.

The Mitsubishi A6M5, 61-120 / N46770 Zeke or Zero-Sen, is one of the most treasured aircraft at The Air Museum and is the only completely original one of its type still flyable in the world.

The Air Museum's collection current aircraft consists of over one hundred and fifty aircraft, of which, thirty are flyable examples of their type. The total number of aircraft continually changes because The Air Museum continually strives to obtain and restore as many historically valuable aircraft as possible. Fighter Rebuilders operates as an on site restoration facility for the Museum and is constantly working on new restoration projects, bringing aircraft back to life in their original flyable condition and others just for static viewing. The flyable aircraft are maintained and flown regularly at airshows around the country, on various motion picture projects and during special flying events held at the Museum. This allows the aircraft to be displayed for future generations in their natural environment - Flight!

The Air Museum currently has the largest displayed collection of World War II Japanese aircraft in the world. Even more significant, is that most of these aircraft are the only surviving ones of their type in the world.

The first aircraft acquired for the museum some fifty years ago was the Mitsubishi J8M-1 Shusui rocket fighter. Only a few prototypes were constructed and this is the only one of its type that survived.

The flyable Mitsubishi A6M5 Zeke or Zero-Sen, as it was called by U.S. Intelligence, is the most authentic one of it's type flying anywhere in the world. It is the only original Nakajima Sakai powered Zero-Sen flying and is the pride of the museum. The museum also has another original Zero-Sen on display but the main wing spare was cut in half, so it's not a flier.

The Fugi Ohka 11 "Baka" Bomb was a rocket powered aircraft designed as a last resort suicide weapon.

The only original Aichi D3A Val dive bomber, like those used during the attack on Pearl Harbor, isn't flyable but is on display. The Museum does have a flyable Vultee BT-15 that was modified for the movie "Toro! Toro! Toro!" to look like a real Val. **(B&W Val photo by Frank Mormillo)**

The Mitsubshi J2M3 Raiden Jack is a sole survivor of its kind. The aircraft is almost complete and original. With a lot of work and money it could be restored to flyable condition.

Two other Japanese aircraft at the museum are the Mitsubishi G6M Betty and the Yokosuka D4Y Suisei Judy, which are displayed in the same condition that they were found in the jungles on a South Pacific Island some forty years after the war.

The museum also had a very rare flyable Nakajima Ki-84 Frank, but the aircraft was sold in the early years to help keep the museum operational.
(Photo by Garald Liang)

After the war, many Japanese and German aircraft were returned to the United States for testing and evaluation. This Mitsubshi J2M3 Raiden or Jack, was one of those aircraft and was fortunate enough to have survived the cutting torch.

The only known Mitsubishi G6M Betty in the world is displayed as it was found in the jungles of New Guinea. The aircraft was recovered only a few years ago, with a number of other Japanese aircraft that had been abandoned after the war. Makes you wonder how many more aircraft are hidden in the jungles on South Pacific Islands?

AIR MUSEUM AIRCRAFT LISTING

Aichi D3A2 Val
Antonov An-2 Colt
Bacham BA349 Natter
Bede BD-5 Vee
Beech SNB / C-45
Bell YP-59A Airacomet
Bell P-39N Airacobra
Bell X-1 replica
Bell X-2 replica
Benson Gryo-Copter
Blaty Oron Hang Glider
Boeing E-75 Stearman (2)
Boeing B-50A Superfortress
Boeing B-17G Flying Fortress
Boeing P-12E
Boeing FB-5
Boeing P-26 Peashooter
Bristol F2B replica (2)
Cessna A-37 Dragonfly
Cessna L-19 Birddog
Chanute 1896 Hang Glider
Convair F-102A Delta Dagger
Convair CV-240
Convair L-13A Grasshopper
Culver PQ-14 / TDC-2
Curtiss 1910 Pusher replica
Curtiss P-40N Warhawk
Curtiss R3C-2 replica
Curtiss C-46 Commando
DeHavilland Vampire MKIII
DeHavilland Vampire MKVI
Deperdussin 1913 Schnieder cup racer replica
Douglas SBD-5 Dauntless
Douglas AD-4 Skyraider
Douglas A-4 Skyhawk
Douglas D558II Sky Rocket
Douglas DC-3 cockpit
Douglas RB-26C Invader
Easy Riser Hang Glider
Fairchild PT-19 Cornell cockpit
Feisher FL-103 Flying Bomb
Fokker DR.1 Triplane replic
Fokker D.VII 1/2 scale replica
Folland T-1
Ford 5-AT Tri-Motor
Fugi Ohka I Baka Bomb

Gnat (2)
Globe KD6D-2 Drone
Gloster Meteor
Grumman F3F-2
Grumman FM-2 Wildcat
Grumman F6F-5 Hellcat
Grumman F8F Bearcat
Grumman F7F-3N Tigercat
Grumman F9F-5P Panther
Grumman F11F-1 Tiger
Grumman F-14A Tomcat
Grumman TBM-3 Avenger
Hanriot HD-1 Scout
Hawker Hunter MKVI
Heinkel He-162A-1 Volksjager
Hiller Helicopter
Hispano Ha 1112 / Bf-109
Horton HO IV Flying Wing Glider
Howard DGA-5 Ike racer replica
Hughes H-1 racer replica

Laister-Kaufman LK-10
Lockheed C-121A Constellation
Lockheed F-104 Starfighter (2)
Lockheed P-38J Lightning
Lockheed L-18 Lodestar
Lockheed C-60 Lodestar
Lockheed T-33A Shooting Star (3)
Lockheed P-80A Shooting Star
Lockheed SR-71 Blackbird cockpit
Macchi M-39 replica
Martin TM-61 Matador
Messerschmitt Bf-109G Gustov
Messerschmitt Me-108B Taifun
Messerschmitt Me-163B Komet replica
Messerschmitt Me-262 Schwalbe
Mikoyan-Gurevich Mig-15 Fagot (3)
Mikoyan-Gurevich Mig-17 Fresco
Miles & Atwood Special replica

Miss Cosmic Wind formula one racer
Mitsubishi A6M5 Zeke 52 Zero (2)
Mitsubishi J2M3 Raiden
Mitsubishi J8M1 Shusui
Mitsubishi G6M Betty
Nieuport 28 replica
North American P-51A Mustang
North American P-51D Mustang (3)
North American F-86A Sabre
North American F-86F Sabre
North American QF-86H Sabre drone
North American QF-100D Super Sabre
North American AT-6 / SNJ Texan
North American FJ3 Fury
North American B-25 Mitchell
North American O-47A
North American T-28B Trojan
North American T-28D Trojan
North American Navion
Northrop F-89J Scorpin
Northrop GAM-54 Crossbow
Northrop N9MB Flying Wing
Northrop MQM-74 Chukar II
Pitts Special
Radioplane RP-5A drone
Radioplane RP-54D drone
Radioplane RP-76B drone
Republic P-47G Thunderbolt
Republic YF-84A Thunderjet
Republic F-84E Thunderjet
Republic F-84F Thunderjet
Republic P-84F Thunderjet

Republic RF-84K Thunderstreak
Republic F-105B Thunderchief
Rick Jet RJ-4
Rider R-4 Firecracker replica
Rider R-6 "8 Ball" midget racer
Rockwell T-2A Buckeye
Rutan Vari-eze
Rutan Quickie Homebuilt
Ryan FR-1 Fireball
Schmidt Commuter Helicopter
Schupal & Nylander Flying Wing
Seversky 2PA/AT-12 Guardsman
Sieman Schuckert D.IV
Sikorsky H-34 Sea Bat
Stearman PT-17 Kadet
Stinson L-5G Sentinal
Sump'n Else #35 midget racer
Supermarine Spitfire PR MKXIX
Supermarine S6B replica
U.S. Navy Bat Missile
Vought F8U-1 Crausder
Vought FG-1D Corsair
Vought BT-13 Valiant
Williams W-7 Stinger
Yakovlev Yak-11 Moose
Yakovlev Yak-19 Max
Yokosuka D4Y Suisei Judy

12 CHINO'S AIR RACING AIRCRAFT

After World War II, military fighter type aircraft were finding their way into civilian hands, some being used as personal business aircraft, while most were just flown for pleasure.

It wasn't long after the war that National Air Racing attracted the use of ex-military aircraft. The years right after the war, 1946 through 1951, saw many aircraft utilized for the Bendix Race, which was flown from Van Nuys, California to Cleveland, Ohio. More than twenty aircraft would enter these long nonstop races. Another race was the Thompson Race, where a course was set up with four pylons in Cleveland, Ohio. The aircraft flew 20 laps, about 300 miles around the course. Races where for groups of aircraft or where only individual types could race.

The early Air Racing pioneers were; Paul Mantz, Tex Johnson, Loen Gray, Dale Fulton, Jack Wollams, Charles Tucker and Thomas Call, just to name a few. They were flying aircraft that had been demilitarized and sold as surplus like, P-51 Mustangs, P-39 Airacobras, P-63 Kingcobras, P-38 Lightnings, FG-1D Corsairs and even A-26 Invaders. Even back then the aircraft were being modified for better performance. Wing tips were being clipped, canopy and cowlings were streamlined for less drag and engines were being modified to produce more horsepower.

The ex-military type aircraft races only lasted a few years, it was after the 1949 race, they were discontinued until 1964, when the first Reno National Air Race was held at Sky Ranch, near Reno, Nevada. This would bring a new breed of pilots to air racing; Darryl Greenamyer, Walt Ohlrich, E.D. Weiner, Ben Hall, Miraslav Slovak, Clay Lacy, Howie Keefe, John Church, Mike Carroll and Stan Hoke, to name a few. The new event at Reno would bring the more highly modified aircraft back to compete in the air racing. These modified aircraft were and still are,

P-51D, SN44-73343 / N5482V is one of the many Mustangs rebuilt at Aero Sport. This aircraft raced at the 1972 Reno National Air Races as #2 "Seattle Miss," but the aircraft was written off in an accident in 1987.

North American P-51D, SN44-74996 / N5410V, "Dago Red," raced as #4, and was owned by Frank Taylor. The aircraft won first place in the Gold 1982 race at Reno. Since then the aircraft has had a number of different owners to including; Bill Destefani, Alan Preston and David Price. (Photo by Philip Wallick)

contining to fly faster and faster each year. Since 1964 the Reno National Air Races have become an annual event and pilots and aircraft continue to make history!

The importance of Chino Airport was not only with Warbirds, but the Chino Airport also has played a major role in modern air racing. Over the years, many air racing aircraft have emerged from the facilities at Chino. Some of these aircraft were stock, while others had aerodynamic modifiactions to improve their performance. There were others strickly built for air racing. Some of these aircraft were the fastest in their class, while others were pilots wishing to experience the thrill and competition of racing. Two of these aircraft would set propeller driven aircraft speed records that still hold today.

The following pages have fragments of information about most of the air racing aircraft that were built or modified in the shops at Chino. All of these aircraft passed through hangars at Chino some time during flying career.

In 1980, Frank Taylor and Bill Destefani built P-51D, SN44-74996 / N5410V / "Dago Red," #4, at Chino. They acquired the expertise and help of Bruce Boland, one of Lockheed Aircraft Company's top aerodynamic engineers, to improve the performance capabilities of the Mustang. The aircraft won the championship the first time it was raced at Reno National Air Race in 1982. (Photo by Gerald Liang)

P-51D, SN44-73149 / N6340T, raced as "Candy Man," #7, in 1976 and was owned by Charlie Beck and Ed Modes. The aircraft currently flies as part of the Fighter Collection in the United Kingdom as G-BTCD and is painted as SN46-3221 "Moose." (Photo by Gerald Laing)

P-51D, SN44-63481 / N6303T, "Diet Rite Cola," raced as #13 in 1970 and later flew as "Miss Gatorade" and was owned and raced by Richard Kestler. The aircraft was totally destroyed in an accident in 1972. (Photo by Gerald Liang)

All of the Mustangs in civilian hands were once operated by a military some where in the world. P-51D, SN44-73973 was once operated by the RCAF and the FA Salvadorena Air Force. Upon its return to the U.S. it was re-registered as N51JC, "Cotton Mouth" after being rebuilt at Aero Sport in 1975. The aircraft flew as C-GCJC in Canada and was later sold to David Price and reverted back to N51JC and has been raced as #49.
(Photo by Gerald Liang)

P-51D, SN44-73140 / N169MD, raced as #71 in 1969. The aircraft would later be sold to an owner in Canada and in 1984, it would return to the U.S. for rebuild at Pioneer Aero as SN44-14151, "Petie 2nd," for warbird collector Doug Arnold in England. (Photo by Gerald Liang)

P-51D, SN45-11558 / N6175C was raced as #39 in 1992. The aircraft is a cavalier conversion and was rebuilt at Aero Sport in 1972 from parts of two other Mustangs. (Photo by Gerald Liang)

P-51D, SN44-73415 / N6526D had a long history of owners, one of which crashed the aircraft in Washington. The aircraft was sold and brought to Aero Sport in 1977 for a complete rebuild. Since then, it has been raced as #45 and #55 -"Pegasus," and currently races as #55 -"Voodoo Chile." (Photo by Gerald Liang)

TF-51D, SN44-84860 / N327DB, "Lady Jo," was raced as #81 in 1995. The aircraft was built at Fighter Rebuilders for Daryl Bond. When raced, the aircraft is piloted by Robbie Patterson. (Photo by Gerald Liang)

North American P-51D, SN44-63865 first showed up at Chino in 1971, registered as N6163U, it was changed after being overhauled at Aero Sport to N51JK, and raced as #99. The aircraft placed 5th in the 1979 Silver and 4th in the 1980 Bronze race at Reno. Up until 1998, the aircraft still wore its, racing colors and had been hangared at Chino.

"The Galloping Ghost" parked on the ramp at the 1946 Cleveland Air race. During the 1950's, the exact whereabouts of the aircraft show it as being exported and utilized by the Israeli Air Force for many years. The serial number SN44-15651 would later resurface on the U.S. registration and the aircraft was sold in 1960. (Photo by Emil Strasser)

In 1972, "Miss Candance" returned to Reno highly modified with a low profile bubble canopy, a new smaller oil cooler air scoop and a new F8F Bearcat propeller. Over the next few years the aircraft still suffered from engine problems and didn't fare too well at the races. (Photo by Emil Strasser)

North American P-51D, SN44-15651 / NX79111, "The Galloping Ghost," #77, was purchased as war surplus with only 1,000 hours on the airframe. Bruce Raymond and Steve Beville would prove to be very competive race pilots, placing first or in the top five, in every race they entered "The Galloping Ghost" in, during the early Air Racing years, 1945-49. (Photo by Emil Strasser)

The aircraft was bought in 1960 by Cliff Cummings, who over the years had it rebuilt and modified a number of times at Aero Sport. "Miss Candance" first raced as a relatively stock P-51D Mustang at the 1969 Reno National Air races. (Photo by Emil Strasser)

In 1970, Cliff Cummings was forced to make a dead stick landing after experiencing an engine failure. The oil cooler dog house was completly riped off the aircraft. This would be the first of many engine failures that "Miss Candance" was plagued with during its career. (Photo by Gerald Liang)

The aircraft was sold and rebuilt again in 1982 with even more modifications to the airframe and wings. Jimmy Leeward renamed his new racer "Specter" and raced with roman numerial #X rather than the number 10.
(Photo by Gerald Liang)

The 1986 racing session would see race #9 still painted bright yellow, but carrying a new name, "The Leeward Ranch Special." (Photo by Gerald Liang)

Cummings sold "Miss Candace" after racng it almost twenty years, after accidents and a number of engines failures made it impractical to maintain. Cliff Cummings sold it to Dave Zeuschel and Dennis Schoenfelder in 1979, who later sold it to Wiley Sanders, who flew the aircraft as #69, "Jeannie," in 1979.
(Photo by Emil Strasser)

Jimmy Leeward returned to Reno with N7911 painted yellow and carrying #9 as a new race number in 1985. (Photo by Emil Strasser)

The aircraft would change again to a bare metal finish, still carrying "Leeward Ranch Special" markings and race as #44. There is no question that North American P-51D, SN44-15651/ N79111 has the most historical air race record of any aircraft currently flying.
(Photo by Gerald Liang)

Painted white and wearing race "Miss RJ," race #5 was modified with a more aerodynamic canopy, streamlined propeller hub and clipped wings for the 1969 Reno Air Race.
(Photo by Emil Strasser)

The first year the aircraft appeared as the "Red Baron" was in 1973, with a new owner, Ed Browning of "The Red Baron Flying Service," of Idaho. In 1974, the "Red Baron" would get its first victory, with Mac McClain flying.
(Photo by Gerald Liang)

1976, saw a few more modifications to the "Red Baron," with the addition of a dorsal fin and even larger tail surface. The dorsal fin proved to be ineffective and was removed the following year. (Photo by Emil Strasser)

A three way partnership between Chuck Hall, Frank Lynott and Charlie Willis was the first ownership of P-51D, SN44-84961 / N7715C, race #5, after the aircraft was surplused from the military. Chuck Hall first started racing the Mustang in 1966, when it was still stock. (Photo by Gerald Liang)

Race #5 was sold to Gunther Balz, in 1971 and returned to Reno painted silver and renamed, "Roto Finish." (Photo by Emil Strasser)

RB-51, 'Red Baron," race #5 during testing at Chino after extensive modifications were made to the airframe and the addition of the Griffin engine with twin contra-rotating three bladed propellers. The larger tail surface is also very visible in this photo. (Photo by Gerald Liang)

The aerodynamic testing and structure analysis for the Red Baron was done by Lockheed Aircraft Engineer, Bruce Boland. Bruce started with a stock P-51D Mustang, SN44-84961 / N7715C, had raced as "Miss RJ" #5. The aircraft was then modified in the Aero Sport shops at Chino into a modified air racing aircraft named, Roto Finish, race #5, which was only raced one year. The same aircraft was then highly modified into the RB-51, Red Baron, which was powered by a Griffin 54 engine built by Dave Zeuschel Racing Engines. It was also equipped with twin contra rotating three bladed propellers.

Well advanced for its day the Red Baron did very well in competition. With Mac McClain at the controls it won first place in the 1977 Gold race at Reno National Air Races. Then in 1978, Darryl Greenamyer won first place in the Gold at Reno and at Mojave. Steve Hinton then flew the Red Baron to first place in the 1979 Miami National Air Race. The aircraft was also raced and won the 1979 Mojave Air Gold Race. At the Reno National Air Race in 1979 the Red Baron set its fastest qualifying speed at 441.9 MPH with race speeds at 415.4 MPH. Then, just after crossing the finish line in second place, the Red Baron experienced an engine failure and crashed into the desert. Steve Hinton was seriously injured, and the aircraft was totally destroyed.

In 1978, Steve Hinton would fly the "Red Baron" to victory at both the Mojave Air Race and the Reno National Air Race. The following year still with Steve Hinton at the controls, the "Red Baron" set a new piston powered airplane speed record at 499 miles per hour. (Photo by Emil Strasser)

No question that P-51D, SN44-74756 / N69QF, "Miss Sis-Q," #33, carried some the most colorful markings put on a Mustang. The aircraft was raced as a stock aircraft by Ken Burnstine, who was later killed while flying this aircraft when it crashed after the 1976 Mojave Air Race. (Photo by Emil Strasser)

In 1973, Ken Burnstine had P-51D, SN44-74502 / N70QF #34 "Miss Foxy Lady" highly modified for use strictly as a race plane by Leroy Penhall at Chino. The aircraft carried a beautiful all black paint scheme. (Photo by Emil Strasser)

After the Mojave Air Race in 1973 "Miss Foxy Lady" was sent to Bruce Gossling of Unlimited Aircraft Limited at Chino for more improvements and a new white paint scheme. The aircraft was sold to John Crocker and was raced as #6 "Sumthin Else." (Photo by Emil Strasser)

In 1976, "Miss Foxy Lady" was sold to John Crocker and repanited white and blue with a new name, "Sumthin Else"and raced as #6. The aircraft was flown to victory in the 1979 Unlimited Class at the Reno National Air Race. The aircraft has since been removed from air racing and undergone a conversion into a TF-51. (Photo by Gerald Liang)

F8F-2, BU122629 / N777L, "Able Cat," first raced in 1969 after being modified at Aero Sport, where a new Wright R-3350 engine and a DC-7 Hamilton Standard propeller were installed. (Photo by Bob Nightingale)

"Able Cat" was first raced in 1969 and appeared in this very drab yellow paint scheme as race #70. (Photo by Gerald Liang)

In 1970, "Able Cat" carried race #77 and was painted a very nice looking yellow and blue. The aircraft suffered minor damage when it ground looped after making an emergency landing, because of engine problems during qualifications. (Photo by Gerald Liang)

"Able Cat" was still raced as #77, the name was changed to "Rare Bear," and became one of the most colorful and popular aircraft on the circuit. Lyle Shelton also set a new world piston engined speed record at 528 MPH in 1989. (Photo by Gerald Liang)

Hawker Sea Fury T.Mk205 VX281 / N8476W "Nuthin Special," raced as #40. The aircraft was owned and flown by Dale Clark and Dennis Firestone. The aircraft was operated from the Sanders facility at Chino. The aircraft was powered by a stock Britsol Centaurus eighteen-cylinder engine with the standard five bladed propeller.

Sea Fury N8487W, raced as #40, was once owned by Richard Drury, who sold it to John Sandburg, who sold it to Dale Clark. The aircraft is currently owned by Wally Fisk of St Paul, Minnesota and flies as "Dragon of Cymru," RN281 / N281L. (Photo by Emil Strasser)

Hawker Sea Fury, FB MK.11 WG567 / N878M, was raced by Sherman Cooper as "Miss Merced," #87 and later as #42 "Super Chief," when owned by Jim Mott. In earlier years, Jim raced a North American AT-6, SN41-34382 / N3274G, "Mis-Chief," #42. (Photo by Gerald Liang)

Just a little too much fuel on start! After many years of being hangared at Chino, Jim Mott sold the Sea Fury. The new owner has had Sanders Aviation, in Northern California completely rebuild the aircraft. To include a new Pratt & Whitney R-3350 powerplant and repaint the aircraft back in "Miss Merced" markings. (Photo by Gerald Liang)

Hawker Sea Fury T MK.20 VZ368 / N20SF, "Dreadnaught," is raced as #8 and owned and flown by Brian and Dennis Sanders. This highly modified Sea Fury has raced since 1983 with a Pratt & Whitney R-4360 engine that produces more than 4,000 horsepower, with a Douglas AD-5 Skyraider propeller. The aircraft won the Gold at the 1983 and 1986 Reno Air Races. The effectiveness and longevity of this Sea Fury has to be considered the most cost-effective race plane still racing. (Photo by Philip Wallick)

Hawker FB MK.11 Sea Fury N19SF, is raced as #19 by Brian and Dennis Sanders. This is the newest aircraft from the Sanders Sea Fury stables and flies with a Pratt & Whitney R-3350 and four bladed propeller. The aircraft has since been painted into Royal Canadian Navy markings BC-1 114, "Argonaught."
(Photo by Gerald Liang)

Hawker Sea Fury T MK.20S VX300 / N924G, was raced as #88, flown by Brian and Dennis Sanders. This aircraft is still powered by the original Bristol Centaurus 3,270 Cubic inch, 18 cylinder, two-row radial air-cooled engine. (Photo by Emil Strasser)

In 1982, the "Super Corsair" made it's initial test flight at Chino after being built at Fighter Rebuilders. Below, shown taxing during the 1983 Reno Air Race. (Both photos by Gerald Liang)

Vought "Super Corsair" - F4U-1D / N31518 raced as #1 and was built by Jim Maloney and Steve Hinton at Fighter Rebuilders in 1982. The aircraft was put together from a fuselage, a hulk airframe and pieces from many different sources. A massive Pratt & Whitney 4360 powered the aircraft, which produced over 4,000 horsepower and used a Douglas AD-5 Skyraider propeller. Bruce Boland, a Lockheed Aerospace Engineer, was called in to do the aerodynamic design and airframe structure modifications.

Over the years, many different pilots flew the Super Corsair at a number of different races around the country. These pilots included, Jim Maloney, Steve Hinton, John Maloney and Kevin Elderidge. The aircraft won the Gold race during the 1985 Reno Air Races.

The Super Corsair was lost after suffering a catastrophic engine failure at the 1995 Phoenix 500 Air Race. The only alternative that Kevin Elderidge had was to bailout. He did so just after the aircraft caught fire, suffering major injuries to his neck and shoulder.

"Super Corsair," while flying to the 1995 Phoenix 500 Air Race this would be one of the last inflight photos of the aircraft before it suffered an engine failure during the race and crashed into the desert. The aircraft was a complete loss. (Photo by Jerry Wilkins)

"Super Corsair," F4U-1D / N31518 raced as #1 with Johnny Maloney flying at the 1990 Texas National Air Race. The massive Pratt & Whitney R-4360 and Douglas AD-1 Skyraider propeller.

It took Fighter Rebuilders almost eight years to build the specially designed race plane for John Sandberg. "Tsunami" / NX39JR raced as #18 and was powered by a beefed up Rolls Royce Merlin with four radically designed Aero Products propellers. Bruce Boland was called on to do the aerodynamic design work on "Tsunami." After a two year construction period, the aircraft emerged with great hopes of being very competitive. The aircraft was raced in 1989, 90, 91 at the Reno National Air Races and is shown above in an author's photo at the 1990 Dallas National Air Race. Skip Holm flew "Tsunami" at the 1991 Reno National Air Race in its new Red Paint scheme. Only a few weeks later the warbird scene would lose another important contributor, when John Sandberg was killed while flying "Tsunami" back to Minneapolis.

(Bottom photo by Gerald Liang)

The "Pond Racer" was a graphite composite aircraft designed by Dick Ratan and owned by Bob Pond. The aircraft was powered by two highly modified 6-cylinder Electramotive VG30 automotive racing engines, which produced more than 1,000-horsepower each. Two four-bladed Hartzell propellers were modified for ground clearance.
(Photographed enroute to the 1991 Reno National Air National Race by Jerry Wilkins)

Though impressive looking, the aircraft never produced the speeds that were expected of it. Tragedy struck during a qualifying run at the 1993 Reno Air Race, when pilot Rick Brickert was killed while attempting to land after an engine failure and fire. (Photo by Gerald Liang)

13 PRIVATE WARBIRD OWNERS

This is what people dream of, owning your own Warbird. There have been and still are, quite a large number of privately owned ex-military aircraft based and operated from the runways at Chino Airport. The oldest structures on the airport are the main four hangars that were built to support the pilot training activities during World War II at CAL-AERO Academy. These same facilities now support a number of different aircraft related operations. Currently Chino Airport is host to the largest number of Warbird restoration companies at one airport than any other airport in the World. Over the years, a number of new hangars where built that currently house many privately owned Warbirds and other aircraft related facilities.

Over the years there have been a variety of Warbirds based in the old and new hangars on Chino airport. The following is as complete a listing as I could acquire of the privately owned warbirds that have operated from the runways at Chino:

Jack Kistler - North American P-51D,
    SN44- 63865 /- N51JK
    This airframe had 3 Kills during WWII
David Webster - North American P-51D,
    SN44-73822 / N117E
Lee Schaller - North American P-51D,
    SN44- 84962 / N9857P
Gerry Miles - North American P-51D,
    SN44-74469 / N7723C
John Marlin - North American P-51D,
    SN44-74458 / N65206

Elmer Ward - North American P-51D,
    SN44-72739 - N44727 "Man O War"/
    Grumman F8F-2/G-59A
    BU122708 / N7701C "Gulfhawk"

Joe Heartney - North American P-51D,
    SN44-84952 / N210D
Al Ashbourne - North American P-51D,
    SN44-73206 / N3751D
Steve Hinton - North American P-51D,
    SN44-84961 / N7715C
Al Conner - North American F-5K, SN44-12840
Daryl Bond - North American T-6
    North American TF-51D,
    SN44-84860 / N327DB

Steve Hinton / Johnny Maloney / Kevin Eldridge - Grumman F8F Bearcat
    BU121614 / N7957C
Harold Beal - Grumman F8F-3
    BU121752 / N2YY
    Douglas A-26K
    SN44-35505 / N4851E
Jim Mott - Hawker FB MK.11 Sea Fury
    WG567 / N878
    North American T-6
    SN41-34382 / N3274G

Tom Nightingale - North American SNJ-4
  BU10116 / N75964
Dick Fields - North American T-6C
  SN88-12150 / N7055H
  Complete project T-6 in restoration
Bill Miller - North American T-6G Texan
Dennis Buehn - North American AT-6D
  SN41-34402 / N3146G

Ray Dieckman - North American AT-6D
  SN41-34653 / N3258G
  DeHavilland DH100/15 Vampire FB MK6
  J-1152 / N152RD
  J-1129 / N4024S
  Hawker Hunter 4060
  Vought FGID Corsair
  BU92433 / NX773RD

Dennis Firestone & Dale Clark - Canadian
  Harvard MK4 RCAF 20257
  Rebuilt as an NA-50 / N98474
Brandon Kunicki - North American SNJ-4,
  BU27253 / N48119
  2 - North American T-6 Texans N?
Chris Fahey - North American AT-6D
  SN42-86505 / N9013A

Erney Banks - North American Harvard
  MK4 RCAF20247 / N1811B
Jack Lowery - North American T-6 Texan
  SN51-14916 / N27817
Ron Thompson - North American T-6 Texan
  SN51-8211 / N432RT
  Stearman / SN75-6419
  Beech D-18 / N44638
Gary Whiteford - North American T-6 Texan
  SN42-17575 / 545GW

Charles Nichols - North American T-6
  SN44-81494 / N7448C
Mike and Dick Wright - North American B-25J
  SN44-30832 / N3155G
L.W. Richards - North American B-25J
  SN44-86891 / N3337G
Al Redick - Douglas B-26C
  On Mark Conversion to B-26K
  SN44-35505 / N4815E

Bob Nightingale - North American
   T-28A SN49-1645 / N99394
   T -28S SN51-3557 / NX85228
   North American T-6 / N81854
   Piper J-3 Cub / N25996
   Consolidated L-5 / N8035H

Bill Melamed - North American T-28C
   BU140576 / N289RD
Brian Kenney - North American T-28D-5
   SN49-1496 / N2496
   Consolidated Vultee OY-1
   BU03902 / N62080
   Beech C-45

Rich Toby - North American T-28A
Wally McDonald - North American T-28A
Tom Johnston - North American T-28A
E.G. Husband - North American T-28A
   SN49-1514 / N9669C
Allen Krosner - North American T-28B
   BU137760 / N128KA

Bud Murphy - North American T-28S
   SN51-3513 / NX9868A

Ron Evans - North American T-28S
   SN51-7533 /N14144

Don Lee - North American T-28
Anton Ostermeier - North American T-28B
        BU138171 / NX171BA
        Folland Gnat T Mk.1, XP514 / N7HY
Roscoe Diehl - North American T-28A,
        SN49-1540 / NX99395,
        Beech C-45 N2833G
        Beech T-34A

Scotty Roberts - North American T-28B
        BU138340 / NX5524L

Walley McDonnell - North American T-28B
        BU138303 / N9671N
Al Grant - North American T-28C
        BU140511 / NX140AG

Tom Neal - North American T-28B
        BU137716 / N415FR
        T-28C BU140053 / N28TN
        Stearman PT-17 & Twin Beech

Pat Nightingale - Stinson 0-68 Sentinel
    In restoration
Doug Damiano - Stinson L-5 Sentinel
    SN214880 - aircraft in restoration
Jeff Person - Cessna 195 SN40-7541 / N9848A

Mike Talbot - Beech T-34A Metor
    SN52-7360 / N44MT
Daryl Bond & Joe Tidewell - Beech T-34A
    Metor N337AR
Dale Medcalf - Beech T-34A Metor
    N4178A
Richard Cunningham - Consolidated Vultee
    BT-13 / N67083
Glen Wilson - Beech T-34A & BT-13 / 15

Sam Stewart - C-47 & Martin VC3A / N404CG
Jim Fredlind - Douglas C-47
Larry Hill - Lockheed Lodestar
Roger Johnson - Beech C-45
Jim Merizan - De Havilland Mosquito
    FB MK VI PZ474 / N9909F
    B MK35 TA639 / N6867C
Joe Haley - Yakovlev Let-C.11 / N2124X
Dale Clark - Yakovlev Let C.11
    Hawker Sea Fury T.MK205 - VX281 /
    N8476W
Sam Davis - Yakovlev Yak-3U / N498SD
Tom Camp - Yak-11 / N18AW
    Grumman FM2 - storage only
Jim Nunn - Grumman FM-2 BU55627 / N7960C

Doug Zeisner - Chinese CJ-6
        SN2132018 /N114DZ
Barry Hancock - Chinese CJ-6
        SN1332010 / N99YK
Robin Scott - Yak-52 / N2255C

Troy Cobb - Cessna L-19 Bird Dog
        SN52-2761 / N305BD
Frank Vanacar - Cessna L-19 Bird Dog
        SN52-2118 / N5199G
Jim Mulvihill - Cessna L-19 Bird Dog
        SN56-2620 / N305AF
Jimmie New - Cessna L-19 Bird Dog
        SN52-3206 / N5182G
Dave Hudson - Cessna L-19
        SN52-1096 / N5073W
Monty Bow - L-19 Project
Mac McCauley - Stearman PT-17
        N450WT
Carter Teeters - Stearman PT-17
        N79997
Jack Davis - Stearman PT-17
George Wildy - Stearman N2S-4 / PT-17
        SN75-2857 / N7122T
Mark Warden - Stearman N2S-4 / N7122T
        Consolidated L-5 Stentinal
        SN51-14916 / N27817
Mike Walsh - Stearman PT-17
        SN75-7721 / N5114N
Willis Hicks - Stearman PT-17
        N62917 & N5085N
Hartley Folstad - Stearman PT-17 - N622SR /
N621SR / N450SR / N450JN / N65263

Dick Breta - Vought F4U-5NL Corsair
  BU124486 / N49068
  Grumman F7F-3 Tigercat
  BU80483 / N6178C
  Douglas AD-4NA Skyraider
  BU126935 /N2088G
  Douglas AD-5W Skyraider
  BU135152 / N65164
  Hawker FB MK.11 Sea Fury
  WM483 / N42SF
Don Hanna - Douglas AD-4NA Skyraider
  BU126959 / N2088V

Some time between 1965 and 1980 there was a Boeing C-97 Stratocrusier, a Fairchild C-82 Packard, a Douglas DC-6 and a Lockheed C-121 Constellation, that were all based and operated from Chino. I am sure that there were more Warbirds that were either privately built or that were based and operated at Chino Airport, but based on the information available, this is as complete a list as could be made.

Over the years and even more so in recent years there have been a variety of privately owned jet type warbirds at Chino, including the, North American F-86 / CL-13, Lockheed T-33 / CL-30, Lockheed F-104, Sabb Draken, Hawker Hunter, Anipodean Fouga, Aero L-29 Delfin, Aero L-39 Albatos, De Havilland DH112 Sea Venom, Folland Gnat, De Havilland DH100 Vampire, BAC Strikemaster, Hispano HA200, Atlas Impala, Mig 15/17, parts of a Douglas A-4 Skyhawk, Vought F-8 Crusader and even a Grumman A-6 Intruder.

KNOWN SURVIVING WARBIRDS WORLDWIDE

	Total Surviors	Total A/C Produced		Total Surviors	Total A/C Produced
BRITISH AIRCRAFT			**UNITED STATES AIRCRAFT**		
Beaufighter	11	5562	Bell P-36	3	317
Beaufort	12	1821	Bell P-39	43	9558
Blenheim	46	3983	Bell P-63	17	3303
Boomerang	16	250	Boeing B-17	51	12731
Firefly	23	658	Boeing B-23	10	38
Gladiator	9	746	Boeing B-29	32	3970
Hudson	8	2000	Boeing Stearman	1000	9000
Hurricane	69	14233	Cessna L-19	300	3399
Landcaster	19	7366	Consolidated B-24	19	18482
Mosquito	32	6439	Consolidated PBY	106	3290
Sea Fury	61	775	Consolidated PB4Y2	8	977
Short Sunderland	7	741	Consolidated OS2U	8	1000
Spitfire	191	22440	Curtiss P-40	79	13779
Swordfish	12	2392	Douglas AD-4 / A-1	53	3180
Tempest	21	800	Douglas SBD	26	5936
Typhoon	3	330	Douglas A-20	15	7385
			Douglas A-26 / B-26	132	2446
GERMAN AIRCRAFT			Grumman J2F Duck	10	645
Fw190	24	21001	Grumman F3F	6	166
He-111 / CASA 2.111	15	7450	Grumman F4F	41	7905
Ju52 Stuka	43	4850	Grumman F6F	28	12275
Ju87	5	5709	Grumman F7F	14	364
Ha 1112	32	264	Grumman F8F	31	1263
Messerchmitt Bf109	31	35000	Grumman TBM	95	9939
Messerchmitt Bf110	6	6050	Lockheed PV-1	44	2000
			Lockheed PV-2	43	1600
JAPANESE AIRCRAFT			Lockheed P-38	28	9923
Aichi D3A2 Val	2	1495	Martin B-26	8	5197
Kawasaki Ki61 Tony	5	3078	North American T-6	780	17631
Mitsubishi A5M Zero	33	10449	North American T-28	351	3001
Mitsubishi G4M Betty	5	2446	North American P-51	301	15686
Nakajima Ki43 Oscar	4	5919	North American B-25	19	18482
Yokasuka D4Y Judy	2	2038	North American P-82	5	270
			Northrop P-63	4	700
SOVIET AICRAFT			Republic P-47	56	15683
Ilyushin IL-2	9	36000	Stinson L-5	100	3000
Lavochkin La-11	4	1182	Vought Corsair	89	12681
Petyakarpov Pe-6	4	11427			
Polikarpov I-16 Rata	12	8644			
Polikarpov I-153	3	3437			
Tupolev Tu-2	27	2500			
Yakovlev Yak-1,3,7,9	11	30000			
Yakovlev Yak-11	51	4566			

There are additional aircraft that could have and maybe should have been added to this list, but accurate information about their total production or their surviving numbers was not available.

14 RELATED OPERATIONS AT CHINO

The following is a list of privately owned companies that operated from, or are currently operating, at the Chino Airport.

AIRCRAFT ELECTRONICS

Chuck Cabe started his aviation career while serving eight years in the U.S. Army. When he returned to the civilian community he worked in Denver, Colorado for an electronics firm. It was after moving to Long Beach, California that Chuck met Frank Sanders while installing a radio system in his T-34. It wasn't long after that, that Chuck was working on all types of Warbirds at "Unlimited Aircraft" at Chino airport. After establishing himself in the Warbird community, Chuck started his own company, "Cabe's Aviation Electronics." Since then Chuck has completely re-wired or installed some sort of electrical component on over two-hundred and sixty different aircraft. The high quality workmanship is easily identifiable and Chuck is recognized as one of the best warbird electrician in the business.

AIRCRAFT FABRICATION "VICTORY AERO"

Mike "LEFTY" McGuckian started working with sheet metal when he was in high school. During his early years in aviation, Lefty worked for Steve Hinton at Fighter Rebuilders doing both mechanics and sheet metal work. It was only a matter of time before Lefty started his own business, "Victory Aero," which continues to fabricate many parts and set pieces for warbird owners. Lefty has worked on many different types of warbird aircraft including, a Beech T-34, Bell P-63, Cessna L-19, Grumman F8F, Lockheed P-38, North American NA-50, P-51, SNJ, T-6 and F-86.

DAVE HANSEN – Another sheet metal and fabrication specialist started building an RV-4 home-built aircraft for himself. During this period, he was selling high dollar inspection equipment to aerospace companies. He found himself enjoying building aircraft so much, he gave up the salesman job and sought work at Pioneer Aero building aircraft components. The first projects that Dave worked on were a P-51D and TF-51 Mustangs that were being constructed for Doug Arnold, a British aircraft collector. He also

Doug Arnold, a British aircraft collector. He also spent 3 1/2 years on Elmer Ward's ill-fated, "Gulfhawk" Bearcat. Other projects that Dave spent a lot of hours on perfecting his trade were building Lockheed T-33 components for Leprino Aviation. While working for the Sanders family, Dave built parts for the following aircraft: the nose section of Kermit Weeks Boeing B-29, a DeHaviland Sea Venom, rebuild of a SF-260 for "Team America," PL-03 improved engine project, heat enclosures for Stearman 450 conversions, a Waco ZPF project, and all the Sanders family Sea Fury's #19, 924, and "Dreadnaught." One of Dave's most recent warbird projects was spending 1 1/2 years assisting in the construction of Sam Davis's Yak-3U. He also just finished a T-6 project for the Confedrate Air Force.

BENT PROP CAFE

The first restaurant on Chino Airport was "Bent Prop Cafe" located in one of the CAL-AERO Academy buildings. Originally opened with an aviation theme in the mid-1960's by owner Hank Servia. The "Bent Prop" became a meeting place for the aviation and the local community. Hank ran the cafe until he was forced to sell it because of health reasons.

FLO'S RESTAURANT

In 1963, Flora Slack moved from her home in Oklahoma and relocated in Southern California to start a new life. She pooled all her money and purchased the Bent Prop cafe, and renamed it "Flo's Restaurant." Her speciality was good old fashion home style cooking and hospitality. Flo's was and still is, one of the more famous attractions at Chino airport. People come from miles, even from other states to eat at Flo's. Flo operated the restaurant for twelve years before she retired in 1975 and sold it.

The current owners, Paul and Donna Hughes purchased Flo's and have maintained both the home style cooking and the hospitality that Flo, herself wished the restaurant to have. Flo's Restaurant is still one of the hot spots at the airport where people come to enjoy the food and airport atmosphere.

VFW

What airport doesn't have one a Veterans of Foreign Wars lounge? Located in one of the original CAL-AERO Academy support buildings. A great place to relax, get something to eat or enjoy a drink with friends, but not before flying!

Don Sakamoto gained his basic knowledge of aircraft propellers during a four year tour in the military. This was during the early 1950's when the military still operated a large number of propeller driven aircraft. He then gained another twenty years experience working at "California Propeller." Instead of retirement, Don started his own company, "Pineapple Propeller, Inc." at Chino Airport in 1990.

Don's shop is a fully certified F.A.A. inspection and repair station. What separates his work from the rest, is, attention to detail and specifications. This results in the highest quality finished propellers being operated.

Don completely built or improved existing propeller repair tooling for his shop. Over the years, Don has repaired every type propeller manufactured for all types of aircraft. Currently he specializes and only repairs, rebuilds and provides final certification for warbird type aircraft propellers. In all his years of operation, no propeller has been returned because of defects in the workmanship. They are the smoothest running propellers being utilized on warbird aircraft today.

AIRCRAFT PAINTING

The process of painting a warbird is much different than a civilian aircraft. The Environmental Protection Agency (EPA) requirements regulate the use of most forms of painting. Some of the rules are, a paint booth has to have a low bypass air filtration system to vent the fumes out and there has to be a two-hour firewall and curtain inside the building to meet fire code standards. Only EPA approved paints can be used and numerous permits are needed for the handling of hazardous waste and materials.

The main difference in the actual painting is the amount of research that goes into the authenticity of the paint scheme and the type of paint used. Not many warbird owners actually use the flat paint that was used on most of the warbirds when they flew with the military. Dupont Imron, Chromabase, PPG, Sikens and Spies-Hecker are the most popular polyurethane type paints used today. But, the paint isn't the most important part of the job, it's the preparation! It takes more time to prep the aircraft than it does to actually paint it.

The process of painting an aircraft consists of stripping the metal with a corrosive chemical stripper. Than a metal brightener is applied and a coat of alodine corrosive protectant is used to preserve the metal. The metal is then sprayed with one or two coats of epoxy paint primer and then the sprayed with two, three, or four coats of the final paint. The last step is to spray one or two coats of a clear-coat to protect the paint and make it last longer. These steps will ensure that the expensive paint job lasts about ten years.

LES HOWARD PAINTING

During the 1960's through the early 1970's, many of the North American P-51 Mustangs that Aero Sport rebuilt were painted by Les Howard Painting facility at Chino Airport. These aircraft were easily identified by the bright colors used, rather than standard drab military colors and markings. These bright paint schemes came along before the warbird movement took off, as owners perferred having pretty toys rather than miltary looking aircraft painted in authentic military markings.

CENTURY AIRCRAFT PAINTING

Tony Corbo owned and operated Century Aircraft Painting from 1979 through 1993. Tony's claim to frame is that he has painted more P-51 Mustangs than any other painter in the U.S., painting both military style paint schemes and any color scheme that the owner wished to have. Some very impressive race aircraft and warbirds that were painted at Century during those years include, Frank Taylor's P-51D "Dago Red", Cliff Cumming's P-51D "Miss Candace," The "Red Baron" Griffin powered RB-51, The Air Museum Planes of Fame "Super Corsair," David Price's Hawker Hurricane, Scotty Roberts North American T-28B, Paul Entrekin's MIG-15, Pascal Mahvi's Canadair CL-30 / T-33 & Douglas A-4 Skyhawk, Chuck Thorton's T-38s, Frank Sander's "Red Knight" Canadair CL-30 / T-33, and Michael Dorn's Lockheed Canadair CL-30 / T-33. Tony remains in the Chino area supplying his aviation knowledge to the warbird community.

CORONA AERO FINISHERS

Kevin Byron and Pat Brass are partners and have been painting aircraft at Chino for over nineteen years. The bread and butter at Corona Aero are the many different types of civilian aircraft and helicopters their shop usually has being painted, some warbird aircraft owners are attracted to having their aircraft painted by the Kevin and Pat, because of the high quality of their work. Over the years, Corona Aero Finishers they have painted quite a few Warbirds including, two PT-17 Stearman, a Grumman F6F Hellcat, Grumman F7F Tigercat & Douglas AD-4N Skyraider, Mikoyan-Gurevich Mig-15, North American P-51 Mustang, a Mitsubishi A6M3 Zero, Supermarine Spitfire MK XVI and a Cessna L-19 Bird Dog.

BRUCE CRANDEL

Bruce has been associated with the Chino airport for over thirty years. Bruce's inspiration was his mother, who painted "Pin-up Girls" on U.S. Navy aircraft at the North Island Naval Air Station in San Diego, California during World War II.

For over 20 years, Bruce Crandel has painted nose art on over 50 different aircraft and has personalized a large number of flying helmets and jackets. He has also painted signs for many different companies' advertising needs.

A Stinson L-5 was the first aircraft that Bruce has painted nose art, since then, some of the aircraft that carry Bruce's art work are, Bob Pond's Grumman F7F Tigercat, Bell P-63 Kingcobra, Grumman F8F Bearcat, The Air Museum Planes of Fame Stinson L-5 Sentinal, Republic P-47G Thunderbolt, Douglas RB-26C Invader, North American P-51D "Spam Can," the "Super Corsair," and the All Coast Race Trailer with a picture of Darryl Bond's TF-51 "Lady Jo." Bruce currently operates out of one of the original CAL-AERO Academy support buildings.

AIRSHOWS AND SPECIAL EVENTS HELD AT CHINO AIRPORT

The early years saw a few small airshows held at Chino, but the first major airshow held on the Chino airport was in 1978 and 1979, they called, "The Gathering of the Eagles."

The Air Museum Planes of Fame began hosting annual airshows in 1980 and discontinued having them in 1992. It was not for the lack of attendance the airshows were cancelled. Economics, insurance and logistics were all determining factors for the cancellation of the airshows. Although, for their Fortieth Year Anniversary, in March of 1997, the Air Museum hosted a large flying display and airshow. The Air Museum does still host monthly events on the first Saturday of each month. These events honor a single type aircraft, or occasion in aviation history.

During one of the early Air Museum airshows the crowd saw this rare formation of aircraft Boeing P-26 Peashooter, Boeing P-12E, Curtiss 0-47, Seversky 2PA / AT-12 Guardian. Each was the only example of its type still flyable in the world. (Photo by Philip Wallick)

The Air Museum - Planes of Fame resumed the annual airshows in 1998, their popularity has always attracted crowds of thousands of local and international spectators.

MOVIES AND TELEVISION

Over the years, many of the aircraft based at Chino, as well as the airport itself, have been utilized for a number of T.V. commercials and Hollywood production movies. These include: "Baa Baa Black Sheep," "Twelve O'Clock High," Abbott and Costello's 1942 movie - "Keep'em Flying," "Best Years of Our Lives," "Rocketman," "Iron Eagle II & III" and "The 1,000 Plane Raid."

THE CHINO "KIDS"

The "CHINO KIDS" was the name given to Jim Maloney and Steve Hinton by Dave Zeuschel. They were the first kids to grow up on the Chino Airport. Mike DeMarino and John and Bill Muszala where also associated with the "Chino Kids" title, but came along later. The combined interest and efforts of these people played an important role in making Chino what it is today.

15 CHINO MISHAPS

Emergency procedures are only as good as the person using them. You shouldn't have to think about them when you need them most during an actual emergency. This is especially true for those people flying any type of high performance warbird. The following is a statement that the U.S. Navy Air Training Command puts on all its pilots training manuals: **"THE BEST TIME TO KNOW PROCEDURES AND THE WORST TIME TO STUDY THEM IS DURING AN EMERGENCY!"** Now, before it's too late, ask yourself, do you know your emergency procedures for your aircraft?

Most people don't like talking about accidents, but anyone who flies needs to continually think of ways to prevent them. There are those non-preventable accidents caused by mechanical failure, but most aviation accidents are caused by human error. Over the years there have been a number of warbird accidents at Chino, or of aircraft based at Chino. Some of them were:

Dick Vartanian, lost a perfectly beautiful stock North American P-51D, SN44-74978 / N74978, in a hangar fire at Shafter Field, near Bakersfield, California. The aircraft was a total loss.

During one of the annual airshows hosted by The Air Museum, Bob Pond crashed his Grumman F6F-5N, BU94473 / N4964W while landing. The aircraft ground looped and rolled over on its back trapping Pond inside until a rescue team could remove him., he suffered only minor injuries. Though the aircraft was damaged, it was later rebuilt at Fighter Rebuilders and still flies today.

While flying a P-51D, John Marlin was in the overhead break to land on Runway 21, when the engine failed due to the lack of fuel. After making a right turn, he realized he couldn't make Runway 21. He then made a left turn, lowered the landing gear, after just clearing Flo's restaurant and missing hangar two, he landed in the dirt off of Runway 26. After adding some fuel, the aircraft was taxied back to its hangar.

A TBM INC., Boeing B-17G, SN44-83546 / N3703G Fire Fighting Tanker, #78, had a main landing gear failure while landing the gear collapsed and the aircraft suffered only minor damage. An Aero Sport maintenance team repaired the aircraft and it returned to fire fighting duty. Years later, the same aircraft would be sold to "Yesterday's Air Force Museum" and later flew in the movie "Memphis Belle."

While flying a PT-19, John Hinton crashed in a field just short of Runway 3, after suffering an engine failure while returning to Chino. Hinton suffered only minor injuries. The aircraft was damaged and later repaired.

Another landing gear problem occurred when the Air Museum's Planes of Fame, North American B-25J, was taxiing and the landing gear collapsed. The mishap was caused when the crew accidentally pulled the landing gear handle up, instead of raising the flap handle. No injuries were sustained and the aircraft was repaired.

Another taxiing accident occurred when David Tallichet was taxiing in his North American P-51D, SN44-63893 / N3333E and collided with a Cessna 210 that was going in the opposite direction on the same taxiway. Luckily, there were no injuries and both aircraft were repaired.

The Air Museum's Grumman F6F-5, BU93879 / N4994V crash landed on the dry lake bed near Lake Elisinore, south of Chino, while attempting an emergency landing. No injuries resulted and the aircraft was repaired and is still flying today.

David Tallichet also had two mishaps with Yesterdays Air Force Republic P-47, SN45-49385 / N47DFD. The first mishap was during takeoff from the Barstow Airport when the engine blower failed and the aircraft crashed at the end of the runway. The aircraft was repaired. The second mishap was another engine failure near Flagstaff, Arizona while enroute to Chino. The pilot suffered only minor injures during both accidents. This P-47D has since been sold and completely restored to military standards by Bill Klaers and Allen Wojak of WestPac Aviation, Rialto, California.

Harold Beal accidentally belly-landed his Grumman F8F, BU121752 / N2YY during a normal landing at Chino Airport. The cause, the pilot just forgot to put the gear handle in the down position. The pilot suffered only minor injures and the aircraft was repaired.

While flying one of Yesterday's Air Forces, Curtiss P-40's, David Tallichet suffered an engine failure while enroute to the Chino airport. The aircraft crash landed just north of March Air Force Base, California only twenty minutes from Chino Airport. The pilot suffered only minor injures and the aircraft was sold and restored by its new owner.

The Air Museum - Planes of Fame Republic P-47G, SN42-25234 / N3395G suffered an engine failure on departure from Naval Air Station Point Mugu after attending an airshow. The pilot suffered only minor injuries and the aircraft was repaired and still flies today.

Wally McDonell's Douglas B-26K, SN43-22649 / N99218 crashed when one of the propellers went into full reverse while on final approach to land at Chino. The aircraft stalled and rolled over crashing upside down only a couple hundred yards from the runway. The aircraft was a complete loss and Walley walked away without injury.

Elmer Ward crashed his beautiful Grumman G-59A Gulfhawk - F8F-5, BU121707 / NL3025 while departing the 1995 Oskhosh Airshow. The aircraft lost power on takeoff, stalled and crashed into a field near the airport.

Elmer suffered only minor injuries, while the aircraft was almost completely destroyed.

Another accident was during the ferry flight of North American P-51D, SN44-74446 / N145D had an engine failure on departure from Chino. The aircraft crashed into a cow pasture just short of Chino's runway. The pilot and passenger suffered only minor injuries, but the aircraft was a write-off. There were a number of causes to this accident, the ferry pilot only had minimal time in high performance propeller driven aircraft. The departure flight was the aircraft's first flight after a seven-year restoration project. No test flights were ever made and the departure was at sunset. This mishap was not just a waste of a Warbird, but was sheer stupidity!

The first lost to the warbird community at Chino, was when Leroy Penhall was a killed in a private aircraft accident while returning from a skiing trip in Colorado.

Dave Zeuschel, of Zeuschel Aircraft Racing Engines, was killed when his F-86 / N86Z suffered a hydraulic failure while flying demonstration at the 1987 Shafter Airshow in central California.

The Warbird community lost one of its pioneers in 1990. Frank Sanders was killed when the "Red Knight" Canadair T-33 crashed while flying near Roswell, New Mexico. Frank was well known to the aviation community for the quality airshow performances he gave and the many contributions he made to aviation.

The Air Museum Planes of Fame suffered its biggest loss in 1981, when Jim Maloney was killed while flying a Ryan PT-22 in Arizona. Jim was one of the original "Chino Kids," older brother to John and son of Ed Maloney.

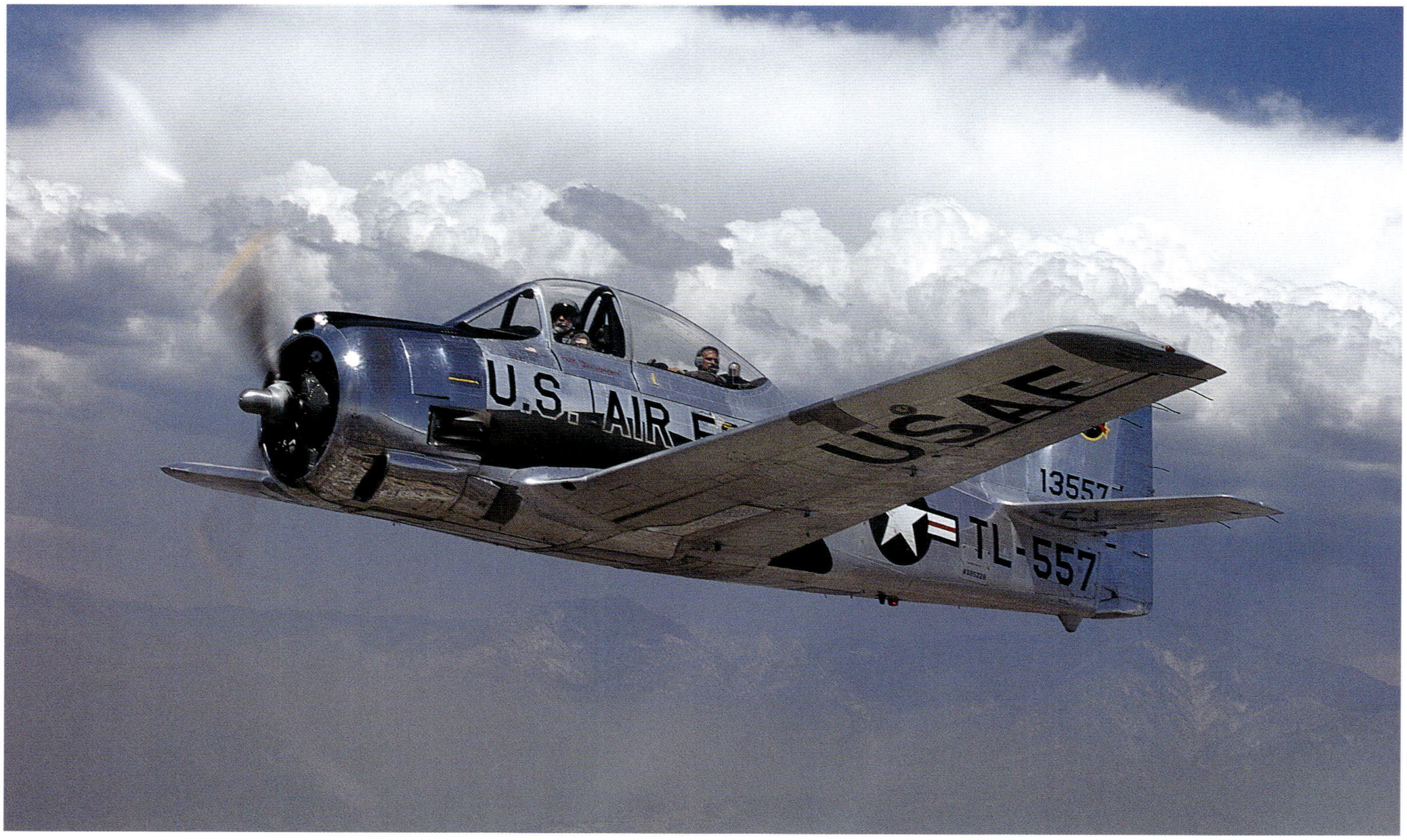
U.S. AIR F
USAF
13557
TL-557

351
WD
VMIT-10

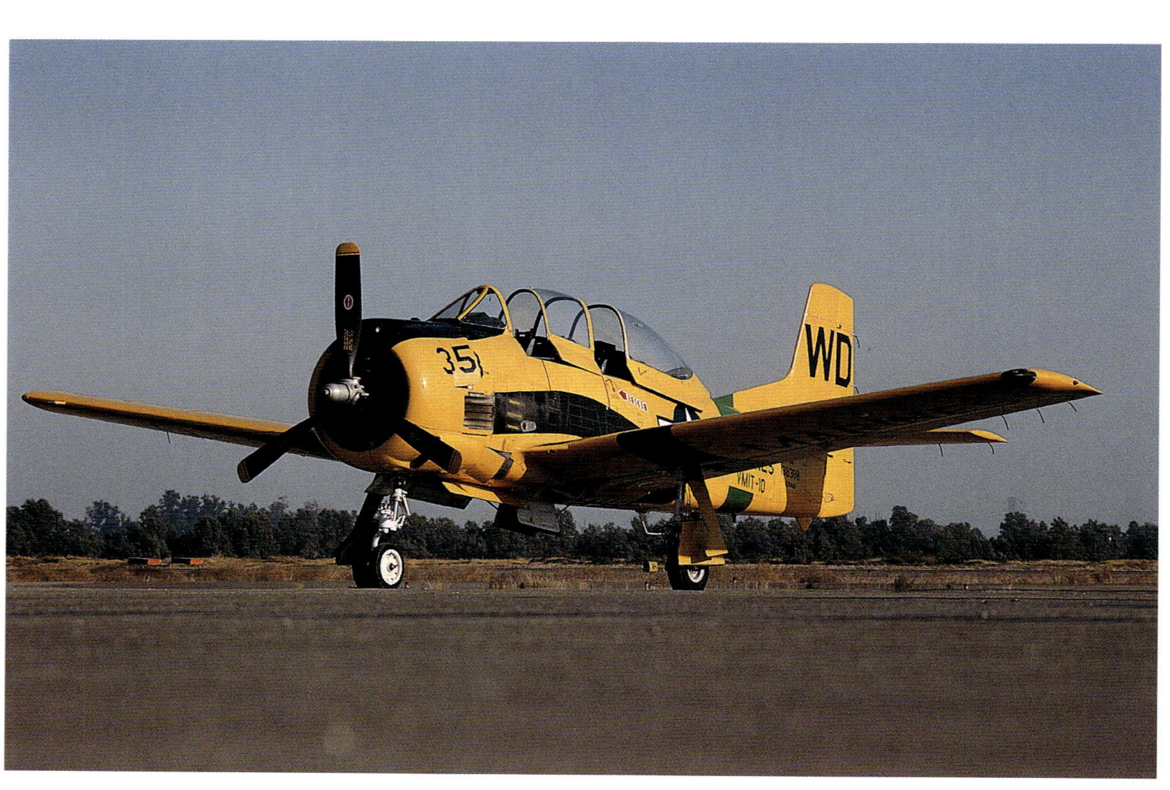